Independent Variables for Optical Surfacing Systems

Haobo Cheng

Independent Variables for Optical Surfacing Systems

Synthesis, Characterization and Application

Springer

Haobo Cheng
Beijing Institute of Technology
Beijing, China

ISBN 978-3-642-45354-0 ISBN 978-3-642-45355-7 (eBook)
DOI 10.1007/978-3-642-45355-7
Springer Heidelberg New York Dordrecht London

Library of Congress Control Number: 2014931851

Printed on acid-free paper

Springer is part of Springer Science+Business Media (www.springer.com)

Contents

Chapter 1
Basic Theory of Optical Surfacing Systems

Abstract This chapter briefly introduces the significance and development of the ultra-precise fabrication of optical lenses and mirrors for current and next-generation optical systems, as well as the basic principle of material removal in deterministic sub-aperture, computer-controlled optical surfacing systems. The assessment criteria for surface quality then are analyzed in detail, covering low-, middle-, and high-frequency ranges.

1.1 Front Matter

1.1.1 The Applications of Optical Segments

After more than 400 years of developments, optics are nowadays everywhere in our daily life and have given rise to great revolutions in our daily activities, such as the camera used to take photos on your travels, the projector used in conferences, three-dimensional (3D) movies at the cinema, the magnifying glass used when reading newspapers, the glasses you wear, and the illuminating systems used at night. Optical technology is also frequently used in industry and production, such as laser cutting and punching, the lithography [1, 2] used to manufacture chips, and machine vision. In the field of metrology, optical technology is used in laser distance measurement, the two-frequency laser interferometer [3, 4] used to measure distance and angle, microscopes used for observing microsubstances, the measurement of surface microstructures, and 3D surface profiling [5, 6]. Optics also have great influence on astro-observation [7–9], geological prospecting, astronautic surveying, weather forecasting, optical communication, infrared night-vision, helmet-mounted or plane displays, medical equipment, inertial confinement fusion [10–12], and laser gyroscopes, etc.

In the twenty-first century, new-generation optical systems are now being developed toward large aperture and large relative aperture, and they have presented increasingly strict specifications on optical material, lightweight design, and surface

H. Cheng, *Independent Variables for Optical Surfacing Systems: Synthesis, Characterization and Application*, DOI 10.1007/978-3-642-45355-7_1,

Fig. 1.1 Twenty-five metre aperture GMT which has seven-circle 8.4 m sub-mirror. *GMT* Giant Magellan Telescope (Reproduced from Cheng 2013)

Table 1.1 GMT parameter of its first mirror

Parameter	Specification	Goal
Radius of curvature	36,000 ± 1.0 mm	±0.3 mm
Conic constant	−0.998286	
Clear aperture	8.365 m	
Off-axis distance	8710 ± 2 mm	±1 mm
Scattering loss L at $\lambda = 500$ nm	<2.0 %	<1.5 %
Actuator correction forces	<50 N root mean square	

accuracy, etc. Meter-class mirrors can be seen everywhere in both ground-based and space telescopes, astronautic cameras, and collimating devices. The relative aperture can reach 1:1 in some cases. Figure 1.1 shows the structure of the famous Giant Magellan Telescope (GMT), which is now being built at the University of Arizona. It comprises seven subsegments, with an aperture of 8.4 m, making it difficult to manufacture and measure. Table 1.1 lists some main parameters of the GMT.

Among the above-mentioned applications, the core point is various optical systems that can control the optical rays to achieve our purpose. The first step in manufacturing these systems is determining how to design them according to desired parameters. Consideration then needs to be given to how to prepare the optical or mechanical segments and how to assemble them together. In past decades, researchers have developed computer-assisted optical designs, and a number of commercial software suites [13, 14] have been used in colleges and companies all over the word. During the design process, the manufacturing capability of current technology must be taken into account. Before the 1990s, people mostly set optical segments as sphere surfaces, which are not so hard to fabricate with high accuracy, but the design scheme may be complex and has a lot of spherical segments to correct aberration and improve image quality.

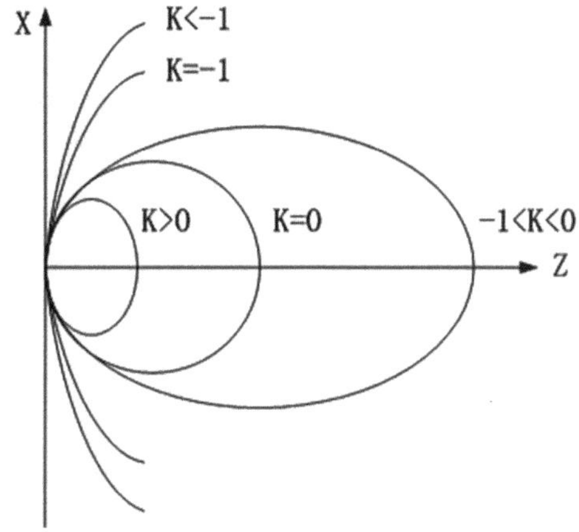

Fig. 1.2 Second-order aspherics with different conic coefficient (Reproduced from Cheng 2013)

1.1.2 Characteristics of Aspherical Surfaces

Broadly speaking, aspheres represent a surface that is different to the sphere surface. It includes symmetrical and nonsymmetrical space surface. The frequently used optical aspherical surface can be generated through rotating a second-order or high-order curve along its symmetry axis. In a rectangular coordinate, it can be defined as:

$$Z = \frac{C(X^2+Y^2)}{1+\sqrt{1-(k+1)C^2(X^2+Y^2)}} + \sum_{i=2}^{n} A_{2i}(X^2+Y^2)^i, \tag{1.1}$$

where C indicates the vertex curvature of the aspherical surface, and k is the conic coefficient. A_{2i} represents the high-order aspherical coefficient. Take second-order aspherics as an example; the conic coefficient can control the surface type, as shown in Eq. (1.2); the correlated schematic diagram is presented in Fig. 1.2.

$$\begin{cases} k > 0 & \text{oblate sphere} \\ k = 0 & \text{sphere} \\ -1 < k < 0 & \text{ellipsoid} \\ k = -1 & \text{paraboloid} \\ k < -1 & \text{hyperboloid} \end{cases} \tag{1.2}$$

Compared with conventional plane and sphere mirrors, aspheric mirrors have more popular advantages. They can increase the freedom of optical system design, improve the image quality, and simultaneously reduce the size and weight of optical systems, which mean less launch cost for space optical systems. However, the fabrication and metrology are more difficult than conventional mirrors because of their non-unique radius of curvature (ROC).

With respect to the fabrication of plane and sphere segments, there have been a lot of mature technologies, such as single-axis or double-axis polishing machines, and annular (i.e. ring) polishing machines. They can produce a perfect plane or sphere surface with accuracy of peak-to-valley (PV) $<1/10\ \lambda$, root mean square (RMS) $< 1/50\ \lambda$ ($\lambda = 632.8$ nm). However, these machines could not work well with an aspherical surface because of their non-unique curvature. Nowadays, the fabrication of aspherical surfaces can be mainly sorted into the following.

- *Material removal method*. This method employs grinding, lapping, or polishing to remove the redundant material or the bulges that are higher than the desired surface. This is the most widely used technology in today's high-accuracy fabrication of optical segments. This work mostly focuses on this point.
- *Material adding method*. This method adds a certain amount of material on the concave pit to build a perfect surface, but is rarely used today due to the instability of newly-added material, especially in a bad working environment.
- *Shape deformation* [15–17]. This method firstly loads stress on the surface to generate a curve, then fabricates it with methods for the plane or sphere surface. After removing the load, the plane or sphere would have an aspheric surface. The fabrication process of the Keck telescope has validated its effectiveness; however, this method needs the help of a mass of stressing fixtures, and some manual intervention to attach and detach the stressing fixtures, which introduces handling risk and does not lend itself easily to automation.

1.1.3 Optical Surfacing Systems: Applications and Developments

Current and next-generation optical systems have presented increasingly critical specifications to optical segments on aspects such as surface form, waviness, roughness and sub-surface damage, etc. The ultra-precise finishing of optical mirrors is complex, iterative, and time-consuming work. In no-computer times, people rely on manual polishing or single-axis machines to produce the perfect surface they need. This work is boring and not quantified, and workers must be careful every second to avoid destroying the surface; therefore, efficiency is quite low and does not satisfy the ever-increasing requirements of the developments of science and technology.

Computers are powerful tools that are used in our daily life and in scientific research. Combined with computer technology, researchers developed various optical surfacing systems that have been widely used all over the world in recent decades. The pioneer who introduced the computer into the fabrication field of optical segments was W. J. Rupp of Itek Optical Systems (1960s); this has been developed into the computer-controlled optical surfacing (CCOS) technology. In terms of a quantified surface from a data form profilometer or interferometer, CCOS technologies utilize a sub-aperture tool to remove redundant materials deterministically, and produce a surface close to our desire.

Fig. 1.3 FSJ3 for the fabrication of large mirrors (Reproduced from Cheng et al. [24])

In the 1970s and 1980s, Jones made a massive effort to promote the development of CCOS [18–22]. He presented the orbital motion model to produce the Gaussian-shape tool influence function (TIF). In 1980, he demonstrated the efficiency of a computer-controlled polisher by experiments on a 1.8 m lightweight mirror. A computer algorithm was written that predicts the surface figure progress for each polishing cycle. A 1.8-m diameter lightweight glass mirror was polished with the computer-controlled polisher for four cycles. The prediction program was used to provide an estimate of expected figure improvement for each cycle. The activity used only 72 h and improved the surface from RMS 0.16 to 0.04 λ. The rapid error correction demonstrates the computer-controlled polisher to be a highly valuable manufacturing tool. Doughty and Smith reported on their computer numerical control (CNC) machine (1987) [23] for the polishing of optical segments with deep asphericity, together with extremely smooth and low damaged surfaces. In their system, three axes were controlled by a microcomputer to drive a very small rotating polishing tool over the rotating aspherical surface. The tool path was controlled to remove material uniformly from the surface or make corrections to the surface profile.

Cheng et al. [24] reported on a multifunctional computer-controlled polishing machine, as shown in Fig. 1.3, which was designed for lapping, polishing, and measuring on-axis or off-axis aspherical surfaces. The manufacturing experiments are carried out by virtue of an off-axis oblate ellipsoid mirror, with rectangular aperture as 770 mm $\times$ 210 mm and centered 127 mm. The accuracy of the aspherical mirror improved from the initial surface form of 17.648 μm RMS to the final form of 0.728 μm RMS after grinding for 200 h.

Since the 1980s, the Steward Observatory has developed stressed lap (as shown in Fig. 1.4) [25, 26] for the fabrication of large astronomical telescope mirrors. The large size, net-like pitch tool could be actively deformed by pressure actuator to

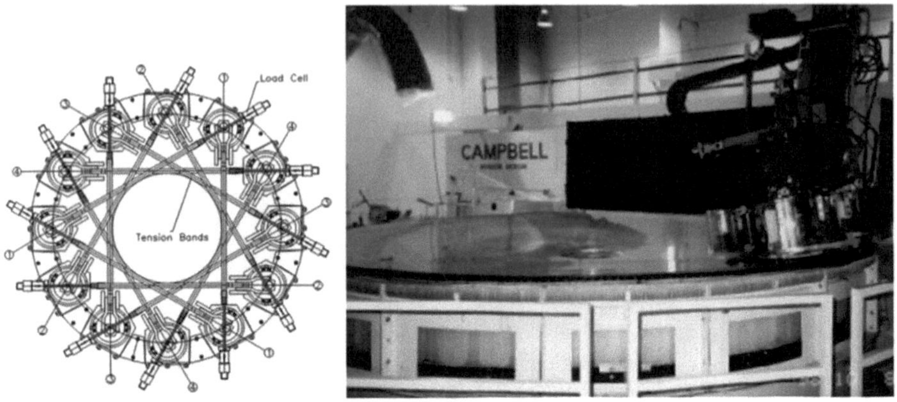

Fig. 1.4 The applications of stressed lap (Reproduced from Cheng 2013)

match the local surface form as the tool moving on the surface. Benefitting from its large size, the tool can effectively smooth the mid-spatial frequency errors. Nowadays, the tool has been successfully used in fabricating large-scale mirrors for the biggest telescopes, such as GMT, etc. In addition, stressed lap has the absolute advantage for the polishing of large aperture optical segments, but the tool is complicated and the cost is high.

Compared with traditional handmade methods, computer-controlled polishing has made a series of great improvements, both in fabrication efficiency and precision. Nevertheless, some inevitable disadvantages are always accompanied, including tool wear, pressure fluctuation, and edge effect, which have promoted the developments of new technologies in recent years.

In the 1990s, Golini et al. [27–29] described the applications of magnetorheological finishing (MRF) in commercial precision optics manufacturing. This method employs a spinning wheel covered by magnetorheological (MR) fluid as the polishing tool. After rheological alteration, MR fluid can produce large, transverse shearing force for material removal when it is presented in the gradient magnetic field. The tool has attracted considerable attention because of its superior features used for material removal, such as high compliance, wear-free, subsurface damage-free, high stability, and sub-nanometer surface roughness. etc. Shorey et al. [30], Kordonski and Gorodkin [31], DeGroote et al. [32], Lambropoulos et al. [33] and Cheng et al. [34] have investigated the material removal mechanism and models of MRF, respectively. Their reports have greatly promoted the widespread application of MRF in the precision fabrication of optical mirrors. In addition, MRF have been successfully commercialized by QED Corporation (as shown in Fig. 1.5).

Electrorheological (ER) polishing was firstly proposed by Kuriyagawa and Syoji [35] in 1999, where ER fluid, together with abrasives (diamond, Green Carbide (GC), or White Aluminium-oxide) in slurry, was used in the polishing of micro-aspheric lenses and metallic dies. The equipment developed for ER fluid-assisted

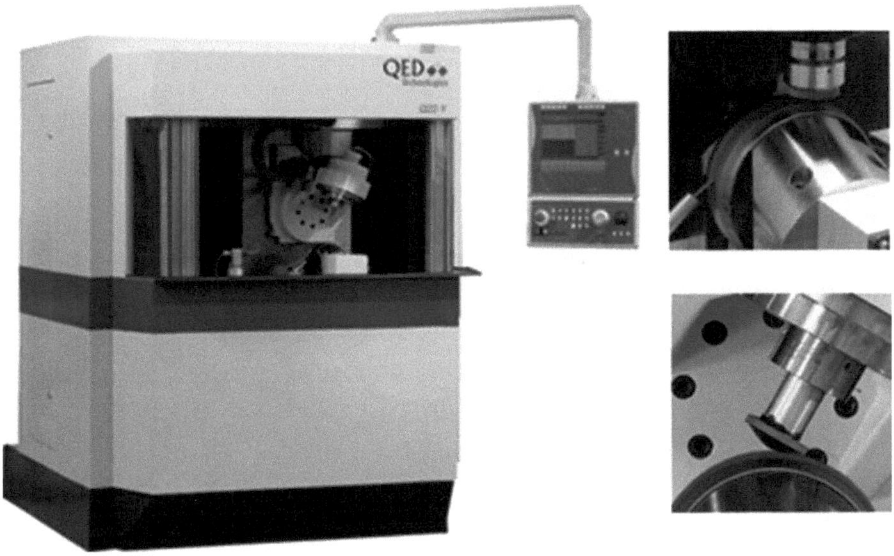

Fig. 1.5 MRF commercialized by QED Corporation. *MRF* magnetorheological finishing (Reproduced from Cheng 2013)

machining [36] involved a spindle with a tungsten carbide tip. While the tip served as an anode for ER polishing, the conductor workpiece was used as a cathode. Kim et al. also investigated ER fluid-assisted polishing, and developed padless ultra-precision polishing using ER fluid [37–39]. A workpiece was put under a mass on a rotating platen, and a high-voltage power supply provided the potential difference between the tool and the platen. Zhang et al. [40, 41] developed a theoretical model for the size of the polishing spot. To prevent short circuiting, Tanak [42] proposed putting a pad between the tool and the workpiece, and Kuriyagawa et al. [43] developed a resin-coated micro-polishing tool using plasma chemical vapour deposition. Figure 1.6 gives the ER finishing setup and the working principle.

Also based on CCOS technology, ion beam figuring (IBF) is a highly deterministic, stable, and non-contact fabrication method, which makes it advantageous over conventional figuring technology such as small-tool finishing [44–51]. IBF (as shown in Fig. 1.7) is performed by directing a broad beam high current ion source (Kaufman type) onto an optic in a carefully controlled manner, removing substrate material by ion beam sputtering, and producing a specified surface figure. Problems that exist in conventional figuring processes, such as the edge effect, tool wear, and pressure variation, disappear naturally during the IBF process. All these advantages distinguish IBF as the final procedure for high-precision optics of various shapes and materials. Carl Zeiss in Germany points out that IBF is the most suitable method for figuring lithography optics [52], and its application sufficiently embodies the machining capability of IBF.

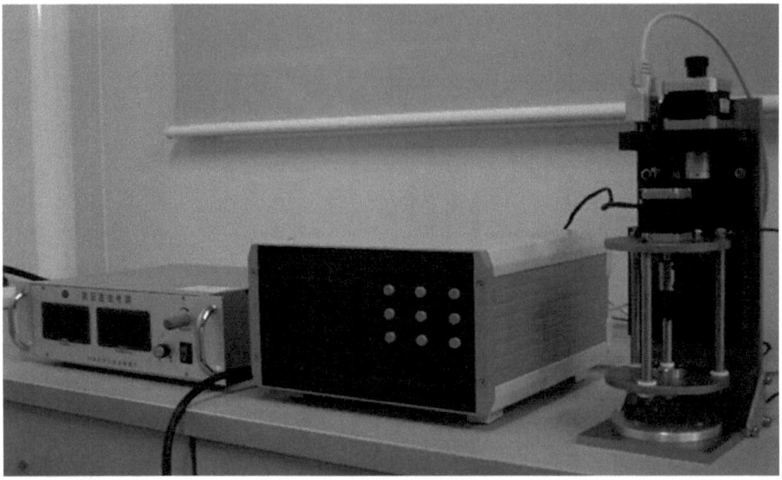

Fig. 1.6 Electrorheological (ER) polishing setup and working principle sketch map

Fig. 1.7 The IBF machine produced by Center Spatial de Liège Laboratory. *IBF* ion beam figuring (Reproduced from Cheng 2013)

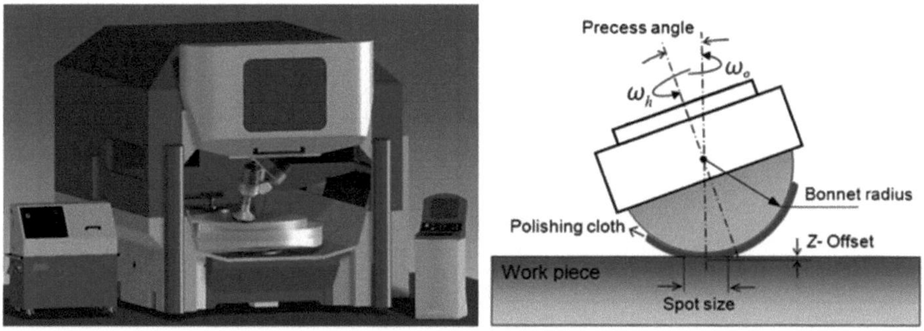

Fig. 1.8 The ABF machine produced by Zeeko Corporation. *ABF* air-bonnet finishing

Walker et al. [53–56] presented their air-bonnet finishing (ABF) tool, which is more compliant (see Fig. 1.8). They also presented a novel motion type called 'Precessions', which drives the tool rotating along its central-axis and processing along the *Z*-axis. ABF is also an even finishing technique as its stable pressure distribution. Through adjusting the air pressure, the ABF tool can produce a large removal rate and a fine surface texture. ABF can be used not only for the finishing of optical mirrors, but is also useful in mould and ceramic polishing. The contact area is always a circle, and it generates Gaussian-shape TIFs. Zeeko Corporation have commercialized it and produced machines with maximum fabrication capability from 200 to 2,400 mm; ABF is now a core portion for the segment fabrication of European Extremely Large Telescope (E-ELT), and the edge effect of the ABF tool was detailed and inspected recently [57, 58].

A new technique, namely MR jet polishing (MJP) [59–61], was initially developed by QED Corporation. This technique can form a stabilized round jet of MR polishing fluid at a distance of several tens of centimeters from the nozzle by magnetic field. Besides, this technique possesses lots of advantages, such as flexible polishing, sub-surface damage-free, polishing tool abrasion-free, deterministic and high-precision polishing in long distance, etc. Kordonski et al. illustrated the basic principles of MJP and researched its removal function. In their studies, a normal mathematical model was built based on the principle that was described as "the rate of polishing of the glass is proportional to the rate at which work is done on each unit area of the glass". They found that the removal function has a W-shaped profile, and a plat surface with high precision was obtained using this method. Moreover, further works about MJP were introduced in the studies of Kordonski et al. [60] and Kordonski and Shorey [61]. Tricard et al. [62] indicated this technique was successful in optical polishing, and especially went well with the components with small volume, thickness, and steep local slopes (Fig. 1.9).

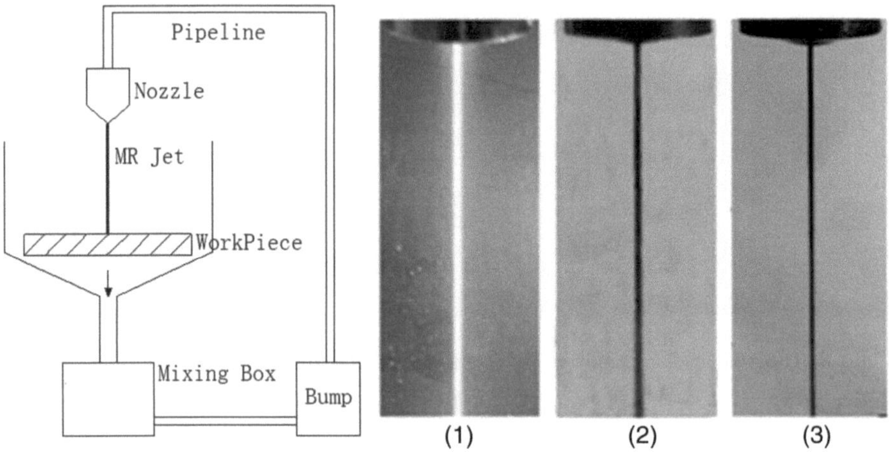

Fig. 1.9 Magnetorheological jet polishing (*MJP*) working principle (*left*) and the jet diagrams (right) with (*1*) abrasive, (*2*) icon and abrasive, (*3*) icon, abrasive and magnetic field. *MJP* magnetorheological jet polishing

1.2 Basic Principle of Deterministic Optical Surfacing Process

1.2.1 Preston Model for Material Removal

Researchers have been always searching a perfect mathematic model for describing the material removal process of optical segments. However, the material removal process is more complex than the machine systems. The existing of chemical and mechanical effects on the working area is hard to be identified and described. Therefore, it is almost impossible to build a perfect model that can completely describe the surfacing fabrication process. We have to make lots of hypotheses, and use a more simplified model to present the surfacing process.

In 1927, Preston gave a classical model used for material removal. He announced that within a large numerical value, the material removal process can be described as a linear function which is proportional to the pressure on the contacting interface and relative velocity between the segment and polishing tool. It can be expressed as:

$$\frac{\mathrm{d}z(x,y)}{\mathrm{d}t} = Kv(x,y)p(x,y) \quad (1.3)$$

where K is a constant depends on material properties of abrasives and workpieces and other process parameters, with the exception of pressure and velocity, $p(x,y)$ is the pressure of point (x, y), and $v(x,y)$ is the relative velocity between the tool and segment at point (x, y).

According to the Preston function, if we know the relative velocity of polishing tools and segments, as well as the pressure distribution model, the material removal can be expressed:

$$\Delta z(x, y) = K \int_0^t v(x, y, t) p(x, y, t) \mathrm{d}t \tag{1.4}$$

This model has been widely accepted as the basic principle used in the CCOS process. It simplifies the complex surfacing process and supplies large convenience for automatic control. However, except for the mechanical cutting effect, the material removal also has close relation with the chemical effect among the optical segments, abrasives, and polishing material. The granularity, concentration and pH value of abrasives, the hardness of optical materials, the motion model of polishing tools, and the temperature all have great influences on the fractal material removal.

1.2.2 Convolution Model of Material Removal Process

The basic idea of CCOS is replacing the handmade method by deterministic optical test and fabrication achieved by the computer's intelligent control. It utilizes a sub-diameter small tool to figure the surface error according to the surface form distribution, and the small tool can well conform to the forms of aspherical surfaces, which can raise the accuracy of surface form.

According to the CCOS principle, the processed material is a convolution of the TIF and the dwell-time map, given as follows

$$E(x, y) = \mathrm{TIF}(x, y) ^{**} T(x, y) = \iint_{\text{path}} \mathrm{TIF}\left(x - x^{'}, y - y^{'}\right) \cdot T\left(x^{'}, y^{'}\right) \mathrm{d}x^{'} \mathrm{d}y^{'} \tag{1.5}$$

where $E(x,y)$ is the amount of material removed on the surface, and $(x - x', y - y')$ is the transformational relation of coordinate systems between TIF and work pieces. $TIF(x,y)$ indicates the TIF, and $T(x,y)$ represents the dwell-time map. Therefore, the calculation of the corresponding dwell time is a deconvolution process according to the anticipant processed material and the fixed TIF.

1.2.3 Linear Shift Invariable System

Actually, the CCOS method described here assumes two basic preconditions:

- The TIF is temporally and spatially stable, which does not change at different dwell points in the whole process.

- When the processing parameters are maintained constant, the relation between the processed material and the processing time must be linear, which means that the processed material in the figuring process is in direct proportion to the dwell time.

Based on the above preconditions, the CCOS process can be seen as such a model: the material can be translated from the motion of the polishing tool; the model is linear and keeps invariant over the polishing duration. This system also can be called linear and shift invariant system.

1.3 Assessment Criteria and Measurements of Optical Surface Quality

As a highly deterministic fabrication process, CCOS needs to know the distribution of surface form exactly. In other words, the CCOS process and the metrology process supplement each other. The fabrication process influences the results of profile metrology, and metrological results guide the fabrication process. Therefore, how to get the distribution map of surface error and how to assess it are the first two steps in deterministic optical surface fabrication.

After 10 years of investigation, researchers have realized that the surface errors should be classified in three frequency stages. Errors with a different frequency range would result in different damages to optical systems, so they have their own assessment criteria and measurements methods. In the 1990s, during the investigation and development of the National Ignition Facility (NIF), the Lawrence Livermore National Laboratory (LLNL) [63–65] divided surface errors into three ranges, which can be expressed as Eq. (1.6).

$$\begin{cases} 33\ \text{mm} < L < 400\,\text{mm},\ f < 0.0303\ \text{mm}^{-1} & \text{low-spatial frequency} \\ 0.12 < L < 33\,\text{mm}, 0.0303\,\text{mm}^{-1} < f < 8.333\,\text{mm}^{-1} & \text{mid-spatial frequency} \\ 0.01\,\text{mm} < L < 0.12\,\text{mm},\ f > 8.333\,\text{mm}^{-1} & \text{high-spatial frequency} \end{cases} \tag{1.6}$$

where L represents the spatial wavelength, and f represents the spatial frequency. Low-spatial frequency errors demand $\lambda/6$ errors for transmitting wavefront, and $\lambda/4$ for reflective wavefront. High-spatial frequency errors should be limited within 0.6 nm RMS. In 2004, the LLNL divided the mid-spatial frequency range into two options: 0.12 mm $< L <$ 2.5 mm and 2.5 mm $< L <$ 33 mm.

1.3.1 Errors in Low-Spatial Frequency Range

Errors in low-spatial frequency range are also called surface form or figure, which can be described by the standard 36-term Zernike polynomial set. They are

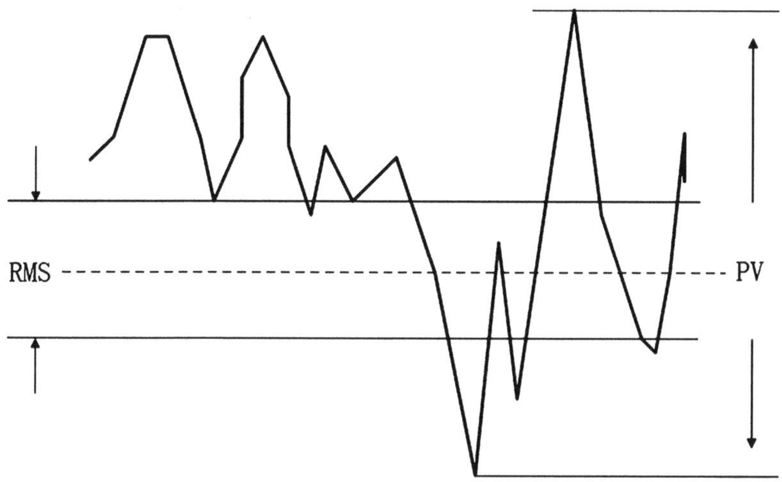

Fig. 1.10 The sketch map of definitions of PV and RMS. *PV* peak-to-valley, *RMS* root mean square

generally assessed by PV and RMS value, as shown in Fig. 1.10. PV defines the difference between maximum and minimum surface height departed from the ideal surface.

$$\mathrm{PV} = \max(z) - \min(z) \tag{1.7}$$

In 1879, after the inspection and research of the image quality of the spectrograph, Rayleigh pointed out that if the maximum difference between practical and ideal wavefront is less than $\lambda/4$, the wavefront could be seen as non-defective. This is known as the Rayleigh criterion. It takes the PV value of wavefront as the assessment criteria. However, nowadays, people have realized that the Rayleigh criterion is not enough to assess the quality of an optical image system. This is because the Rayleigh criterion only pays close attention to the two special points of the wavefront. It ignores the influence of other points. Two optical systems may have different image capability, while possessing the same PV value. In other words, a good PV value does not mean better image quality.

RMS defines the root mean square departure between practical and ideal wavefront, which serves as an important criterion to assess the quality of the wavefront.

$$\mathrm{RMS} = \sqrt{\sum_{i=1}^{n} (z - \bar{z})^2 / n} \tag{1.8}$$

where n defines the total sampling number, and $\bar{z}$ represents the mean value of the surface form dataset. RMS takes all the points into calculation, so it is more significative than a PV value when assessing image quality of an optical system.

The definition of RMS seems not as straightforward as PV. A. Marechal investigated the relationship between the RMS value of the wavefront and central brightness, and concluded that when RMS $\leq \lambda/14$, the Strehl ratio would be greater than 0.8. This is also referred to as the Marechal criterion.

In 1999, Lawson et al. [66], of the LLNL, presented the RMS gradient to specify the low-spatial frequency errors. To meet the Stockpile Stewardship Program (SSP) requirement with its goal of 500 TW inside a 250 μm diameter focal spot, the system should strengthen its focusability. They considered that the wavefront RMS gradient was the most closely tied to the focusability of a laser system. The magnitude of the local wavefront gradient is analogous to the concept of ray angles in geometric optics. The distribution of gradients determines the basic focal spot characteristics. Using this formalism, one can calculate an approximate RMS gradient specification for individual optics in a laser system required to produce a specified spot size. According to their analysis, an RMS gradient of 7.5 nm/cm is adequate to produce 500 TW inside of a 250 μm focal spot at the NIF-designed operating point.

1.3.2 Errors in Mid-spatial Frequency Range

Mid-spatial frequency errors are also referred to as waviness, ripple, and quilting, which can be important but are not explicitly covered by such traditional assessment specifications. Power spectral density (PSD) is a mathematical quantity that defines the spectral content of a signal. The PSD describes how a signal's power is distributed in frequency. In optics, it has been used for some time in a number of applications—notably for specifying high-frequency surface quality to quantify scattering properties. More recently, optical engineers have started to apply PSD metrics to quantify the full spectrum of surface and wavefront quality in optical systems. One of the driving applications for these types of specifications was the NIF at the LLNL, along with a few notable other applications.

PSD has been designated as the preferred quantity for specifying the information in the frequency domain for surface errors. It can be calculated from the surface form measured by an optical or mechanical profiler. The inspection bandwidth limit is dominated by the data density measured by profilers. PSD can be obtained in both integral and digitized form. If assuming $z(x)$ is the profile height in one-dimension as a function of distance x, its finite-length Fourier transform may be obtained from Eq. (1.9):

$$Z(k) = \int_0^L z(x)\exp(-ikx)\mathrm{d}x \tag{1.9}$$

where k is the wave number. Realistically, most profilers give discrete datum, therefore the calculation of PSD should be transformed into digitized form. The equivalent of Eq. (1.9) could be expressed as Eq. (1.10).

$$Z(m) = \Delta x \sum_{n=0}^{N-1} z(n) \exp(-i2\pi mn/N), \; -N/2 < m < N/2 \tag{1.10}$$

where N denotes the data number of the measured profile, and Δx is the spaced interval of sampling points. Total sampling length: $L = N\Delta x$. According to the Nyquist sampling theorem, the frequency range can distinguish locates from $f_{\min} = 1/L$ to $f_{\max} = 1/(2\Delta x)$. The spatial frequency vector $f_m = m/L$. Thus, the wave number is digitized as $k = 2\pi f_m$. The definition of PSD is then expressed as Eq. (1.11) and digitized as Eq. (1.12):

$$\mathrm{PSD} = \frac{|Z(k)|^2}{L} \tag{1.11}$$

$$\mathrm{PSD}(\nu_m) = \frac{\Delta x}{N} \left| \sum_{n=0}^{N-1} z(n) \exp(-i2\pi mn/N) \right|^2 \quad -N/2 < m < N/2 \tag{1.12}$$

where ν_m represents spatial frequency. Attention is paid so that, a PSD curve calculated from a one-dimensional (1D) profile is exceedingly noisy. For smoothing the PSD profile, the average PSD of multilines or 1D curve translated from a two-dimensional (2D) PSD map are usually adopted to stand for the whole surface.

Average PSD curve can be obtained as Eq. (1.13), where p represents the number of sampling lines.

$$\mathrm{PSD}_{\mathrm{avg}} = \sum_{i=1}^{p} \mathrm{PSD}_i / k \tag{1.13}$$

2D PSD can be defined as:

$$\mathrm{PSD}(f_m, f_n) = \frac{\Delta x \Delta y}{MN} \left| \sum_{k=0}^{M-1} \sum_{j=0}^{N-1} z(k,j) \exp[-i2\pi(mk/M + nj/N)] \right|^2 \tag{1.14}$$

In Matlab, 2D Fourier transform is expressed as:

$$\mathit{fft2}(z) = \sum_{k=0}^{M-1} \sum_{j=0}^{N-1} z(k,j) \exp[-i2\pi(mk/M + nj/N)] \tag{1.15}$$

Then, also in Matlab, 2D PSD can be written as:

$$\mathrm{PSD} = \frac{\Delta x \Delta y}{MN} \mathit{fft2}(z) \tag{1.16}$$

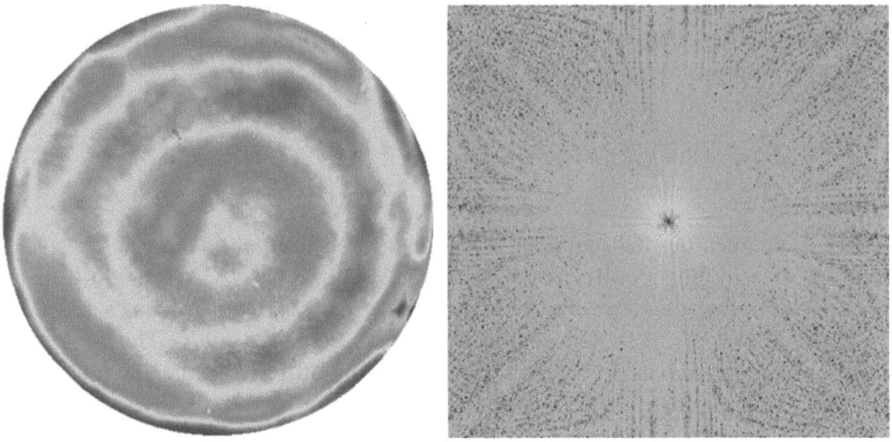

Fig. 1.11 The surface form of 100 mm plane segment (*left*); its 2D PSD map (*right*). *2D* two-dimensional, *PSD* power spectral density

2D PSD is united by the quadruplicate of length: $\text{nm}^2{\cdot}\text{mm}^2$. The 2D PSD map is often short of clearance, therefore we also integrate it along the *X* or *Y* direction to get the 1D PSD curve, as shown Eq. (1.17).

$$\begin{cases} \text{PSD}\,(f_m) = \sum\limits_{f_n=f_{n1}}^{f_{n2}} \text{PSD}(f_m,f_n)f_n \\ \text{PSD}\,(f_n) = \sum\limits_{f_m=f_{m1}}^{\nu_{m2}} \text{PSD}(f_m,f_n)f_m \end{cases} \tag{1.17}$$

The PSD can also be used as a specification unto itself, but it is typical to utilize the PSD to compute the RMS error for given spatial frequency bands. The RMS error is the area underneath the PSD curve. Thus, the RMS over a set of spatial frequencies is simply computed with an integral or sum. The equations used to do this for a discrete and integrated 1D PSD are:

$$\text{RMS} = \sqrt{\sum_{n_{\text{low}}}^{n_{\text{high}}} \text{PSD}(\nu)\Delta f} \tag{1.18}$$

$$\text{RMS} = \sqrt{\int_{f_1}^{f_2} \text{PSD}(f)\text{d}f} \tag{1.19}$$

where $f_1 \leq f \leq f_2$ is the frequency band of interest. Note that the 2D case is simply a double integral (or double sum in the discrete case).

The significance of a PSD curve is illustrated by an example. Figure 1.11 is the surface form of 100 mm plane segment and its 2D PSD map. Figure 1.12a shows the

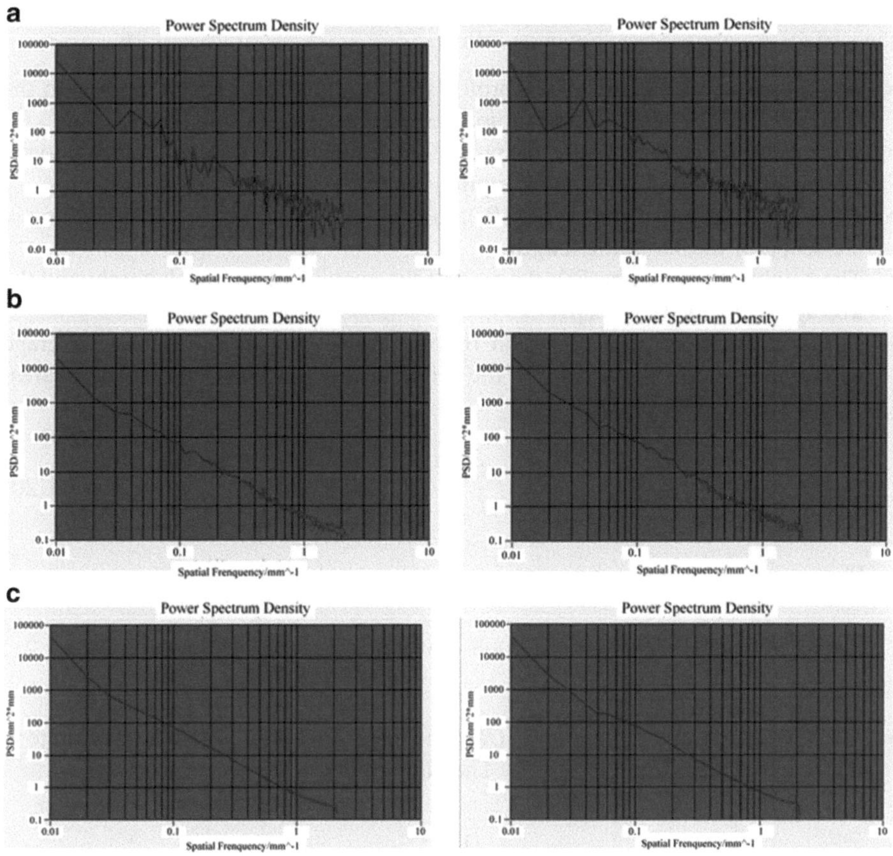

Fig. 1.12 (**a**) The 1D PSD along *X* (*left*) and *Y* (*right*) direction; (**b**) the 1D average PSD along *X* (*left*) and *Y* (*right*) direction; (**c**) the 1D PSD integrated from 2D PSD along *X* (*left*) and *Y* (*right*) direction. *1D* one-dimensional, *PSD* power spectral density, *2D* two-dimensional

1D PSD along the *X* (left) and *Y* (right) direction; Fig. 1.12b shows the 1D average PSD along the *X* (left) and *Y* (right) direction, from which we can see the average PSD curve is much smoother than 1D PSD; Fig. 1.12c shows the 1D PSD integrated from 2D PSD along the *X* (left) and *Y* (right) direction.

1.3.3 Errors in High-Spatial Frequency Range

High-spatial frequency errors are commonly referred to as finish and are quantified using RMS roughness. It is a parameter that is used to evaluate various small machining marks of the surface. The parameters regarding the height character of the surface roughness are shown as follows:

- The highest peak height of the profile (R_p): the distance from the peak to the center line in the scope of the sampling length.
- The maximum deep valley of the profile (R_m): the distance from the valley to the center line in the scope of the sampling length.
- The highest height of the profile (R_y): the distance from the peak to the valley in the scope of the sampling length.
- The height of ten points (R_z): the sum of the average of five peak height and the average of five valley deep.
- The arithmetic average deviation of the profile (R_a): the arithmetic average deviation of setovers absolute in the scope of the sampling length. The expression is shown as follows

$$R_a = \frac{1}{l}\int_0^l |y(x)|\mathrm{d}x \tag{1.20}$$

or

$$R_a = \frac{1}{n}\sum_{i=1}^{n} |y_i| \tag{1.21}$$

- The RMS deviation of the profile (R_q): the RMS deviation of setovers absolute in the scope of the sampling length. The equation is

$$R_q = \sqrt{\frac{1}{l}\int_0^l y^2(x)\mathrm{d}x} \tag{1.22}$$

or

$$R_q = \sqrt{\frac{1}{n}\sum_{i=1}^{n} |y_i|^2} \tag{1.23}$$

References

1. Lowisch M, Kuerz P, Conradi O, Wittich G, Seitz W, Kaiser W (2013) Optics for ASML's NXE: 3300B platform. In: SPIE advanced lithography, 2013. International Society for Optics and Photonics, Bellingham, pp 86791H–86791H–86799
2. Matsuyama T, Ishiyama T, Omura Y (2004) Nikon projection lens update. In: Microlithography. International Society for Optics and Photonics, Bellingham, pp 730–741
3. Donati S, Giuliani G, Merlo S (1995) Laser diode feedback interferometer for measurement of displacements without ambiguity. IEEE J Quant Electron 31(1):113–119

4. Iwasawa K, Iwama A, Mitsui K (2004) Development of a measuring method for several types of programmed tool paths for NC machine tools using a laser displacement interferometer and a rotary encoder. Precis Eng 28(4):399–408
5. Harasaki A, Schmit J, Wyant JC (2000) Improved vertical-scanning interferometry. Appl Opt 39(13):2107–2115
6. Neal RM, Wyant JC (2006) Polarization phase-shifting point-diffraction interferometer. Appl Opt 45(15):3463–3476
7. Johns M, Angel JRP, Shectman S, Bernstein R, Fabricant DG, McCarthy P, Phillips M (2004) Status of the Giant Magellan Telescope (GMT) project. In: Proceedings of SPIE, Bellingham, 2004, pp 441–453
8. Zimmerman DC (2010) Feasibility studies for the alignment of the Thirty Meter Telescope. Appl Opt 49(18):3485–3498
9. Gilmore G (2008) European Extremely Large Telescope: some history, and the scientific community's preferences for wavelength. In: Extremely Large Telescopes: which wavelengths? Retirement symposium for Arne Ardeberg, 2008. International Society for Optics and Photonics, Bellingham, pp 698607–698608
10. Moses EI, Campbell JH, Stolz CJ, Wuest CR (2003) The National Ignition Facility: the world's largest optics and laser system. In: High-power lasers and applications. International Society for Optics and Photonics, Bellingham, pp 1–15
11. Andre ML (1997) Status of the LMJ project. In: Second international conference on solid state lasers for application to ICF. International Society for Optics and Photonics, Bellingham, pp 38–42
12. Peng H, Zhang XM, Wei X, Zheng W, Jing F, Sui Z, Fan D, Lin Z (1999) Status of the SG-III solid state laser project. In: Third international conference on solid state lasers for application to inertial confinement fusion. International Society for Optics and Photonics, Bellingham, pp 25–33
13. Radiant Zemax. http://www.radiantzemax.com/en
14. Synopsys—optical solutions. http://optics.synopsys.com/codev/
15. Lubliner J, Nelson JE (1980) Stressed mirror polishing. 1: a technique for producing nonaxisymmetric mirrors. Appl Opt 19(14):2332–2340
16. Pepi JW (1990) Test and theoretical comparisons for bending and springing of the Keck segmented ten meter telescope. In: The Hague '90, 12–16 Apr 1990. International Society for Optics and Photonics, Bellingham, pp 275–287
17. Rupp V (1965) The development of optical surfaces during the grinding process. Appl Opt 4(6):743–748
18. Jones RA (1977) Optimization of computer controlled polishing. Appl Opt 16(1):218–224
19. Jones RA (1980) Computer controlled polisher demonstration. Appl Opt 19(12):2072–2076
20. Jones RA (1986) Computer-controlled optical surfacing with orbital tool motion. Opt Eng 25(6):256785
21. Jones RA, Rupp WJ (1991) Rapid optical fabrication with computer-controlled optical surfacing. Opt Eng 30(12):1962–1968
22. Jones RA (1995) Computer simulation of smoothing during computer-controlled optical polishing. Appl Opt 34(7):1162–1169
23. Doughty G, Smith J (1987) Microcomputer-controlled polishing machine for very smooth and deep aspherical surfaces. Appl Opt 26(12):2421–2426
24. Cheng H-B, Feng Z-J, Cheng K, Wang Y-W (2005) Design of a six-axis high precision machine tool and its application in machining aspherical optical mirrors. Int J Mach Tool Manuf 45(9):1085–1094
25. Martin HM, Anderson DS, Angel JRP, Nagel RH, West SC, Young R (1990) Progress in the stressed-lap polishing of a 1.8-mf/1 mirror. In: Astronomy '90, Tucson AZ, 11–16 Feb 1990. International Society for Optics and Photonics, Bellingham, pp 682–690
26. Anderson DS, Angel JRP, Burge JH, Davison WB, DeRigne ST, Hille B, Ketelsen DA, Kittrell W, Martin HM, Nagel RH (1992) Stressed-lap polishing of 3.5-mf/1.5 and 1.8-mf/1.0 mirrors. In: 8th international symposium on gas flow and chemical lasers. International Society for Optics and Photonics, Bellingham, pp 260–269

27. Jacobs SD, Golini D, Hsu Y, Puchebner BE, Strafford D, Prokhorov IV, Fess EM, Pietrowski D, Kordonski WI (1995) Magnetorheological finishing: a deterministic process for optics manufacturing. In: International conferences on optical fabrication and testing and applications of optical holography. International Society for Optics and Photonics, Bellingham, pp 372–382
28. Golini D, Jacobs SD, Kordonski V, Dumas P (1997) Precision optics fabrication using magnetorheological finishing. In: SPIE's international symposium on advanced materials for optics and precision structure. International Society for Optical Engineering, San Diego, California, pp 251–274
29. Golini D, Kordonski WI, Dumas P, Hogan SJ (1999) Magnetorheological finishing (MRF) in commercial precision optics manufacturing. In: SPIE's international symposium on optical science, engineering, and instrumentation. International Society for Optics and Photonics, Bellingham, pp 80–91
30. Shorey AB, Jacobs SD, Kordonski WI, Gans RF (2001) Experiments and observations regarding the mechanisms of glass removal in magnetorheological finishing. Appl Opt 40 (1):20–33
31. Kordonski W, Gorodkin S (2011) Material removal in magnetorheological finishing of optics. Appl Opt 50(14):1984–1994
32. DeGroote JE, Marino AE, Wilson JP, Bishop AL, Lambropoulos JC, Jacobs SD (2007) Removal rate model for magnetorheological finishing of glass. Appl Opt 46(32):7927–7941
33. Lambropoulos JC, Miao C, Jacobs SD (2010) Magnetic field effects on shear and normal stresses in magnetorheological finishing. Opt Express 18(19):19713–19723
34. Cheng H, Yam Y, Wang Y (2009) Experimentation on MR fluid using a 2-axis wheel tool. J Mater Process Technol 209(12):5254–5261
35. Kuriyagawa T, Syoji K (1999) Development of electrorheological fluid assisted machining for 3-dimensional small parts. J Jpn Soc Precis Eng 65:145–149
36. Kuriyagawa T, Saeki M, Syoji K (2001) Study of electrorheological fluid-assisted ultra-precision polishing for 3-dimensional small parts. In: Euspen: European Society for Precision Engineering and Nanotechnology. International conference, pp 738–741
37. Kim W, Lee S, Kim Y, Lee E (2003) The electromechanical principle of electrorheological fluid-assisted polishing. Int J Mach Tool Manuf 43(1):81–88
38. Kim W-B, Park S-J, Min B-K, Lee S-J (2004) Surface finishing technique for small parts using dielectrophoretic effects of abrasive particles. J Mater Process Technol 147(3):377–384
39. Kim W, Min B-K, Lee S (2004) Development of a padless ultraprecision polishing method using electrorheological fluid. J Mater Process Technol 155:1293–1299
40. Zhang L, Kuriyagawa T, Kaku T, Zhao J (2005) Investigation into electrorheological fluid-assisted polishing. Int J Mach Tool Manuf 45(12):1461–1467
41. Zhang L, Zhao Y-W, He X-S, Kuriyagawa T (2008) An investigation of effective area in electrorheological fluid-assisted polishing of tungsten carbide. Int J Mach Tool Manuf 48(3):295–306
42. Tanaka T (2007) A study of basic characteristics of polishing using particle-type electro-rheological fluid. Key Eng Mater 329:201–206
43. Kuriyagawa T, Yan JW, Yoshihara N, Kaku T, Abiko K, Mikami Y, Noguchi M (2007) Development of a resin-coated micro polishing tool by plasma CVD method-electrorheological fluid-assisted polishing of Tungsten carbide micro aspherical molding dies. Key Eng Mater 329:213–218
44. Schroeder J, Dieselman H, Douglass J (1971) Technical feasibility of figuring optical surfaces by ion polishing. Appl Opt 10(2):295–299
45. Wilson S, McNeil J (1987) Neutral ion beam figuring of large optical surfaces. In: 31st annual technical symposium. International Society for Optics and Photonics, Bellingham, pp 320–324
46. Wilson S, Reicher DW, McNeil JR (1989) Surface figuring using neutral ion beams. In: 32nd annual technical symposium. International Society for Optics and Photonics, Bellingham, pp 74–81
47. Allen LN, Keim RE (1989) An ion figuring system for large optic fabrication. In: 33rd annual technical symposium. International Society for Optics and Photonics, Bellingham, pp 33–50

48. Drueding TW, Fawcett SC, Wilson SR, Bifano TG (1995) Ion beam figuring of small optical components. Opt Eng 34(12):3565–3571
49. Allen LN, Romig HW (1990) Demonstration of an ion-figuring process. In: San Dieg-DL tentative. International Society for Optics and Photonics, Bellingham, pp 22–33
50. Shanbhag PM, Feinberg MR, Sandri G, Horenstein MN, Bifano TG (2000) Ion-beam machining of millimeter scale optics. Appl Opt 39(4):599–611
51. Allen LN, Hannon JJ, Wambach Jr RW (1992) Final surface error correction of an off-axis aspheric petal by ion figuring. In: San Diego-DL tentative. International Society for Optics and Photonics, Bellingham, pp 190–200
52. Weiser M (2009) Ion beam figuring for lithography optics. Nucl Instrum Methods Phys Res Sect B: Beam Interact Mater Atoms 267(8):1390–1393
53. Walker DD, Brooks D, King A, Freeman R, Morton R, McCavana G, Kim S-W (2003) The 'Precessions' tooling for polishing and figuring flat, spherical and aspheric surfaces. Opt Express 11(8):958–964
54. Walker DD, Freeman R, Morton R, McCavana G, Beaucamp A (2006) Use of the 'Precessions' TM process for prepolishing and correcting 2D & 2½D form
55. Walker D, Beaucamp A, Doubrovski V, Dunn C, Evans R, Freeman R, McCavana G, Morton R, Riley D, Simms J (2006) Commissioning of the first precessions 1.2-m CNC polishing machines for large optics. In: Optics and photonics. International Society for Optics and Photonics, Bellingham, pp 62880P-62880P-62888
56. Yu G, Walker D, Li H (2012) Implementing a grolishing process in Zeeko IRP machines. Appl Opt 51(27):6637–6640
57. Walker D, Yu G, Li H, Messelink W, Evans R, Beaucamp A (2012) Edges in CNC polishing: from mirror-segments towards semiconductors, paper 1: edges on processing the global surface. Opt Express 20(18):19787–19798
58. Li H, Walker D, Yu G, Sayle A, Messelink W, Evans R, Beaucamp A (2013) Edge control in CNC polishing, paper 2: simulation and validation of tool influence functions on edges. Opt Express 21(1):370–381
59. Kordonski W, Shorey AB, Sekeres A (2004) New magnetically assisted finishing method: material removal with magnetorheological fluid jet. In: Optical science and technology, SPIE's 48th annual meeting. International Society for Optics and Photonics, Bellingham, pp 107–114
60. Kordonski WI, Shorey AB, Tricard M (2006) Magnetorheological jet (MR Jet™) finishing technology. J Fluids Eng 128(1):20–26
61. Kordonski W, Shorey A (2007) Magnetorheological (MR) jet finishing technology. J Intell Mater Syst Struct 18(12):1127–1130
62. Tricard M, Kordonski W, Shorey A, Evans C (2006) Magnetorheological jet finishing of conformal, freeform and steep concave optics. CIRP Ann Manuf Technol 55(1):309–312
63. Lawson JK, Aikens DM, English Jr RE, Wolfe CR (1996) Power spectral density specifications for high-power laser systems. In: Optical instrumentation & systems design. International Society for Optics and Photonics, Bellingham, pp 345–356
64. Lawson JK, Aikens DM, English Jr RE, Whistler WT, House W, Nichols MA (1999) Surface figure and roughness tolerances for NIF optics and the interpretation of the gradient, PV wavefront, and RMS specifications. In: SPIE's international symposium on optical science, engineering, and instrumentation. International Society for Optics and Photonics, Bellingham, pp 510–517
65. Campbell JH, Hawley-Fedder RA, Stolz CJ, Menapace JA, Borden MR, Whitman PK, Yu J, Runkel MJ, Riley MO, Feit MD (2004) NIF optical materials and fabrication technologies: an overview. In: Lasers and applications in science and engineering. International Society for Optics and Photonics, Bellingham, pp 84–101
66. Lawson JK, Auerbach JM, English Jr RE, Henesian MA, Hunt JT, Sacks RA, Trenholme JB, Williams WH, Shoup III MJ, Kelly JH (1999) NIF optical specifications: the importance of the RMS gradient. In: Third International Conference on Solid State Lasers for Application to Inertial Confinement Fusion. International Society for Optics and Photonics, Bellingham, pp 336–343

Chapter 2
Tool Influence Functions

Abstract This chapter investigates the tool influence functions (TIFs) and analyzes relevant assessment criteria from the aspect of removal capability. Then, with respect to different optical surfacing systems, TIF theoretical models were constructed and validated by experiments. A series of key parameters were optimized to confirm the preferable TIF for deterministic finishing processes.

2.1 Front Matter for Tool Influence Functions

As mentioned in Chap. 1, the tool influence function (TIF) serves as an important and indispensable variable in the optical surfacing system. It represents the material removal produced by a polishing tool in a unit of time. In the practical polishing process, the TIF would be taken into the calculation of dwell time. Its peak removal rate (PRR) and shape would generate a huge influence on the magnitude and distribution of the dwell-time map.

Based on the Preston function presented in Chap. 1, the theoretical model of TIFs can be built through the corresponding motion model and pressure distribution model. According to the Fourier transform theory, we can conclude two important equations:

$$\widetilde{\mathrm{TIF}} = \int_{-\infty}^{\infty} \mathrm{TIF}(x, y)\ \mathrm{d}x\mathrm{d}y \tag{2.1}$$

$$\mathrm{TIF} = \int_{-\infty}^{\infty} \widetilde{\mathrm{TIF}}(\varepsilon, \eta)\ \mathrm{d}\varepsilon\mathrm{d}\eta \tag{2.2}$$

Equation (2.1) indicates that the direct current (DC) response of polishing systems is constructed by the integration of TIF in the space domain. The integration value is always non-negative because of the non-negativity of TIFs. We can obtain uniform material on the surface if uniform polishing is conducted on the surface. Equation (2.2) illustrates that a TIF with zero central removal can remove a

H. Cheng, *Independent Variables for Optical Surfacing Systems: Synthesis, Characterization and Application*, DOI 10.1007/978-3-642-45355-7_2,

certain period of surface errors but leave more mid–high spatial frequency errors. If the TIF has central peak removal, the surface error can be convergent one-by-one with increasing polishing times.

Thus, from the above analysis, the basic characteristics of TIFs can be concluded as:

- A TIF should be a rotational-symmetric and smooth function.
- A TIF has central peak removal and decreases as the radius increases.
- A TIF has no material removal when the distance exceeds the maximum radius.
- The slope of a TIF at the center and edge regions should be zero.

Since the 1970s, people have taken more than 40 years to find an ideal polishing tool with perfect TIFs. New tools with different work principles emerge every few years. Their purpose is only to improve the performance of material removal on removal rate, roughness, free-edge effect, non-subsurface damage, and stability. Removal rate directly affects the fabrication efficiency, which is referred to as the primary purpose to some extent. The roughness performance is also significant; it influences the distribution of high-spatial frequency errors, which has strict demands in x-ray and laser gyroscope systems. The ideal polishing tool should also have no subsurface damage to ensure that no damage exists on the subsurface of optical segments, as this damage would diminish the long-term stability of optical systems and the coating quality, image performance, and laser-induced damage threshold of optical elements. Contact-type polishing tools remove material by relying on contact pressure, which results in an edge effect when the tool dwells in the edge region of the segment. Stability also serves as an indispensable target, as the convergence rate of surface form would largely decrease if the stability of the TIF is low. The ideal TIF should always remain invariant during the entire fabrication.

2.2 The Assessment Criteria of a TIF

2.2.1 Peak and Volume Removal Rate

A TIF represents the distribution of the material removal rate across the size of the polishing tool in a unit of time. It is often characterized by two parameters; the first is the PRR, which is the maximum removal rate, as presented in Eq. (2.3).

$$\mathrm{PRR} = \max(\mathrm{TIF}) \tag{2.3}$$

The other is the volume removal rate (VRR), which is the sum of all values in the data array multiplied by the cell scale squared, as described in Eq. (2.4). If the shape remains invariant, the VRR would be linear proportional to the increasing PRR.

$$\mathrm{VRR} = \sum_{i=1}^{m}\sum_{j=1}^{n}\mathrm{TIF}(i,j)\cdot\Delta x\cdot\Delta y \tag{2.4}$$

2.2.2 Central-Peak Factor for Removal Capability

Generally, a TIF possessing central peak material removal has a more powerful capability to figure small errors of a surface form. To describe it quantitatively, a central-peak factor is defined as:

$$F_{\text{cpf}} = \frac{\int_{-r/4}^{r/4} \text{TIF}(r)\text{d}r}{\int_{-r}^{r} \text{TIF}(r)\text{d}r} \tag{2.5}$$

where r represents the half-size of the TIF. A large central-peak factor means the TIF is highly centralized and has more power to figure the surface form.

2.2.3 Cutoff Frequency for Removal Capability

The removal capability of a TIF for localized residual errors has great significance for high fabrication accuracy of optical surfaces. From this view, the best TIF should be a pulsing function that which could correct any localized residual errors. However, we could not obtain a pulsing TIF in any case. If a TIF performs closer to a pulsing function, it has much higher removal capability for localized residual errors. For a certain size TIF, the shape has a large influence on the removal capability. From the frequency domain, the normalized amplitude frequency spectrum of a TIF can be expressed as:

$$\text{TIF}_F(w) = \text{FFT}(\text{TIF}_n(r)) \tag{2.6}$$

where FFT represents Fast Fourier Transform, and $\text{TIF}_n(r)$ represents the normalized TIF. An example of a normalized amplitude frequency spectrum curve is shown in Fig. 2.1.

Figure 2.1a displays a typical TIF. Figure 2.1b is the corresponding normalized amplitude frequency spectrum curve, which quickly drops down to near-zero. For a specific spatial frequency, a higher normalized amplitude represents a higher removal capability for localized small errors.

The cutoff frequency represents the specific spatial frequency where the normalized amplitude spectrum curve reduces to 0.05 (point A shown in Fig. 2.1b); we consider that the TIF could not generate a valid correction for the localized error whose spatial frequency exceeds the cutoff frequency. For a TIF with a certain size, the cutoff frequency could evaluate the removal capability for localized errors. A higher cutoff frequency means higher removal capability for localized small errors.

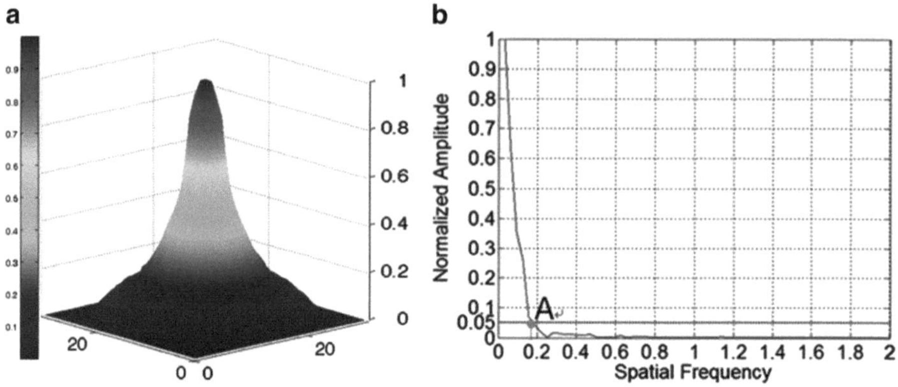

Fig. 2.1 An example of normalized amplitude frequency spectrum curve. (**a**) The typical TIF; (**b**) its corresponding normalized amplitude frequency spectrum curve. *TIF* tool influence function

2.3 TIF Model Constructions and Optimizations

2.3.1 Computer-Controlled Polishing with Small Tools

2.3.1.1 System Descriptions and TIF Model Construction

Computer-controlled polishing with small tools is regarded as the premier computer-controlled optical surfacing (CCOS) technology which has been investigated in detail in the US since 1970 [1, 2] and then been developed widely all over the world. This technique breaks the problems of traditional handmade and single-axis machines in both fabrication efficiency and surface precision, and especially devotes to the fabrication of large aspheric mirrors. China has been pursuing this advanced fabricating technique since 1990. After 20 years of development, several generation systems have been developed.

Figure 2.2 shows an overall structure photograph and movement schematic view of the computer numerical control (CNC) machining system developed in the 1990s. The system is arranged as a gantry structure—two pillars, a crossbeam and a base form the main frame of the machine. Taking machining and measuring stability and performance into consideration, Chinese Tai Shan Qing (TSQ) marble blocks, as the selected material, are shaped into parts of the machine frame. Based on the idea of deterministic manufacture, the tool's self-rotation axis should always trace the normal line of the aspheric surface across the whole working area in order to achieve the required accuracy with high material removal efficiency. Therefore, the system is designed to be controlled on six axes: (1) the tool's self-rotation around its axis at a rotational speed of W_2; (2) the tool's swing around the X-axis; (3) the tool's motion in the Y-axis; (4) the tool's motion along its axis; (5) the turntable's rotation around its axis at a rotational speed of W_1; and (6) the turntable's motion in X direction. Furthermore, a profiler equipped with dual touch-trigger probe (Heidenhain Co., Ltd, Germany) is installed in the system for on-machine profile measurement during the grinding process, which is also an innovating issue of the system.

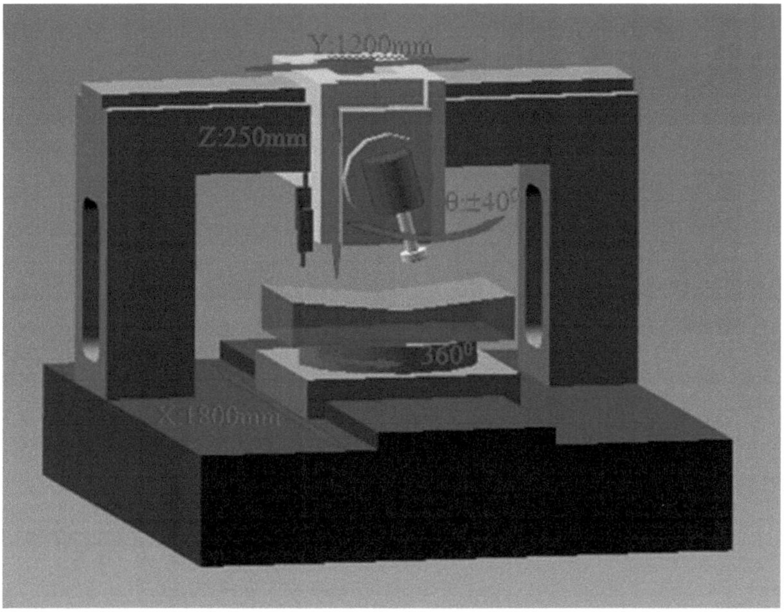

Fig. 2.2 Movement schematic view of the computer numerical control (CNC) machining system (Reproduced from Cheng et al. [7])

Table 2.1 Functions and specification of the machining system

Number of working axials	Stroke in *X* direction	Stroke in *Y* direction	Stretch of tool in its axis	Online testing	Range of swing angle	Surfacing off-axial optics
6	1.8 m	1.2 m	0.25 m	Contact	±40°	Yes

The system combined the faculties of grinding, fine grinding, polishing, and on-machine profile measuring. The sub-aperture tool is driven with two motors. One is for the tool's swing around the X-axis, and the other is for the self-rotation of the tool around its axis. Meanwhile, a workpiece held on the turntable is rotated at the speed of W_1 under the polar coordinate (for the workpiece shaped with rotary symmetry), or moved straight in X direction under the Cartesian coordinate (for the workpiece shaped with non-rotary symmetry, i.e. rectangular shape). In order for the tool axis to always be the normal line of the aspheric surface, the concurrent control for the tool swing angle θ_i and the swing center position in Y direction is made available. The contact pressure between the tool and the workpiece is applied using an air cylinder mounted on the tool axis. Thus, a good fit between the tool and the workpiece is maintained as the tool moves along the aspheric surface. During machining operation, the loose abrasive is injected into the working area on the workpiece surface. The functions and specifications of the developed machining system are shown in Table 2.1. The profiler equipped with two touch-trigger

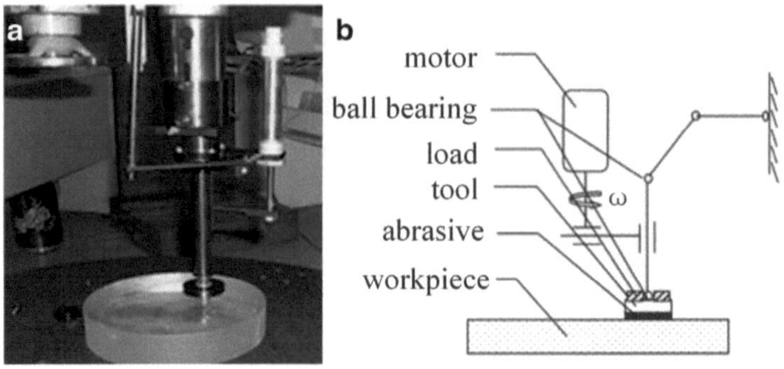

Fig. 2.3 Motion pictorial view. (**a**) The tool planar motion model; (**b**) its motion schematic view of the tool

Fig. 2.4 Sketch map of the tool's motion

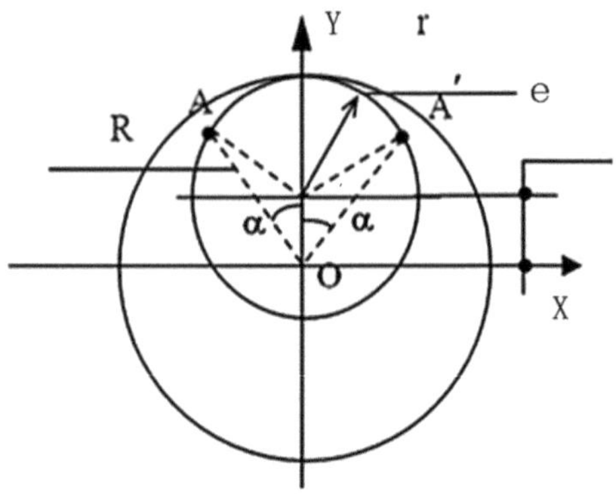

Heidenhain digital length probe assemblies are mounted on a translation stage, using a granite straight edge with high-precision linearity as the reference to reduce the linearity error of the ball slide.

The optimum TIF configuration is determined in order to produce a profile with a central peak and a fairly rapid decrease to zero. A method for generating the desired tool removal profiles was determined to be the epicyclic motion—dual rotations by Jones in 1977 [3]. The tool would rotate around its center at a rapid rate, while the entire mechanism would revolve around a secondary center at a slower rate. Here, a novel tool is designed to move in planar model enlightened from the traditional hand polishing. Figure 2.3a, b present a photograph and motion schematic view of the tool; its motion principle is different to the one in the Jones paper [3]. At the working start position, the tool would rotate around its center at a proper rate from position $+\alpha$ to $-\alpha$ in a clockwise direction driven by a servo motor, to achieve a good tool move orbit, there is often an off-centering distance between the axis of motor and tool, while a draw rod would draw back the tool rotate from position $-\alpha$ to $+\alpha$ in an anticlockwise direction; thus, the to-and-fro motion would be continued by virtue of the reciprocator.

In Fig. 2.4, r is the radius of the tool, e is the off-centering distance, during one machining cycle, the covered grinding or polishing area should be a circle with a radius of $r + e$ and centered as point O, for a random point A located in the

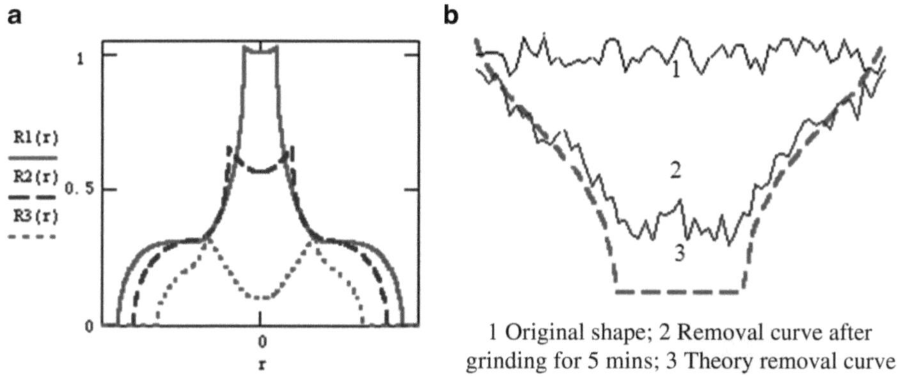

Fig. 2.5 (**a**) Simulation on removal rate with different *e*/*r*; (**b**) experimental result (Reproduced from Cheng et al. [7])

machining area from point O is distance R, the angle with Y direction is α. Therefore, the relative acting angle should be:

$$2\alpha = \begin{cases} 2\arccos\left(\dfrac{R^2 + e^2 - r^2}{2\text{Re}}\right) & r - e < R \le r + e \\ 2\pi & 0 \le R \le r - e \end{cases} \tag{2.7}$$

According to the Preston hypothesis, the material removed under the grinding or polishing tool is given by

$$R(r) = KP\int_0^t V\mathrm{d}t \tag{2.8}$$

where K is the constant that depends on the tool and the workpiece materials and the grinding or polishing compounds, etc., P is the pressure at point A, V is the relative velocity between the tool and point on the workpiece, which is a constant equal to $w \times e$ in one machining cycle, from Eqs. (2.7) and (2.8).

$$\begin{aligned} R(r) &= KP\int_0^t (we)\frac{\mathrm{d}\alpha}{w} \\ &= \begin{cases} 2eKP\arccos\left(\dfrac{R^2 + e^2 - r^2}{2\text{Re}}\right) & r - e < R \le r + e \\ 2\pi eKP & 0 \le R \le r - e \end{cases} \end{aligned} \tag{2.9}$$

To find out a set of optimum technical parameters, we designed three sets of curves (see Fig. 2.5a). $R1(r)$, $R2(r)$ and $R3(r)$ are achieved when e/r equals 0.72,

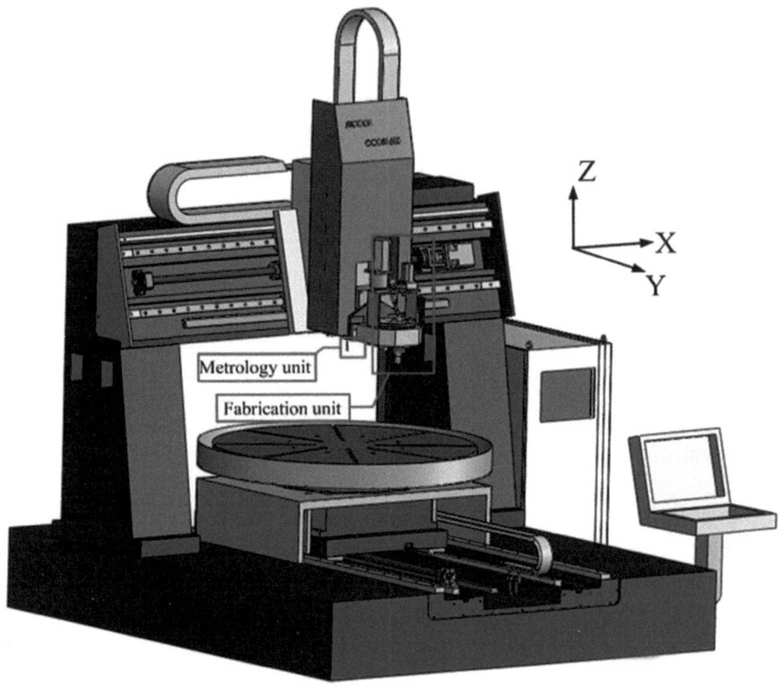

Fig. 2.6 Schematic illustration of JR-1800

0.85, and 1.00, respectively. It is easy to observe that the TIF curve $R1(r)$ with a higher central peak would result in successful machining. Choosing $e/r = 0.72$ as an optimum parameter, at a relative rotational speed $w = 20$ rpm (revolutions per minute), a grinding experiment is performed on a K9 glass for 5 min. The results shown in Fig. 2.5b also verify the selected optimum parameter has a proper removal feature.

Another typical system was built in 2012 (see Fig. 2.6), namely JR-1800. This system is a multifunctional and compound machine system with seven CNC axes. The system adopts a bridge structure on a base marble of 3, 200 mm × 4, 000 mm × 620 mm (weight: ~20 t), which supplies a highly stable fabrication and metrology environment. JR-1800 comprises five parts: a cross *XY*-axis (travel distance: 1, 840 mm × 2, 096 mm), a separate *Z*-axis (travel distance: 603 mm), a rotating turntable (diameter: 1,800 mm), a changeable profile metrology unit, and a fabrication unit. The location uncertainties of *XYZ*-axis are controlled within 5 μm in their travel distance, and linearity is less than 2 μm in any 1,000 mm range.

The machine is capable of producing ultra-precise surfaces for a various of optical materials (e.g. silicon carbide (SiC), fused silica, Zerodur, etc.) and surface forms (plane, sphere, axisymmetric or off-axis aspherics). The maximal fabrication and metrology range reaches up to 1,800 mm. Compared with single-purpose machines, JR-1800 integrates fixed abrasive micro-lapping, loose abrasive polishing,

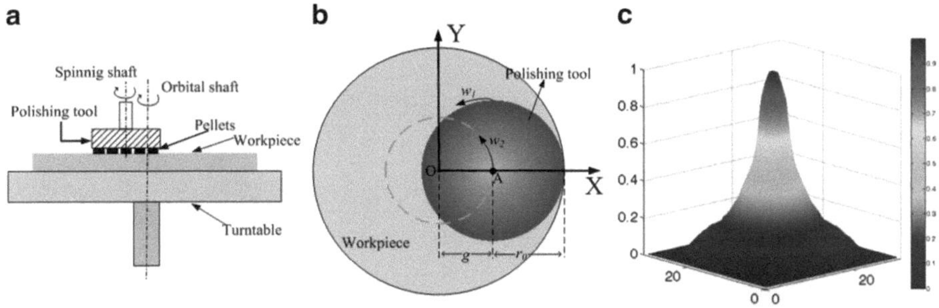

Fig. 2.7 Planetary motion model in deterministic lapping and polishing

and online metrology. It features advantages of (1) shorter fabrication or metrology duration; (2) diminishing the risk when carrying workpieces from one place to another; and (3) largely saving cost and floor space.

It is widely accepted that Gaussian-shape TIF is preferable in deterministic lapping/polishing. To achieve this characteristic, the fabrication unit employs a planetary motion model, as shown in Fig. 2.7a, b. The polishing pad (with semi-diameter r_0) executes self-rotation with respect to its spindle. The spinning velocity w_1 reaches up to 1,000 rpm. Simultaneously, the pad makes an orbital motion (orbital velocity w_2) with changeable orbital radius g. A typical Gaussian-shape TIF of JR-1800 is shown in Fig. 2.7c. With the exception of the spinning and orbital axes, the fabrication unit also employs a tilt axis to ensure that the spindle remains normal to curved surfaces.

This eccentric transmission of spinning is a tricky problem because high rigidity and torque are required for lapping/polishing of hard material. The conventional soft axis system is incompetent to transmit large torque, and lacks stability. We presented a retractable universal joint (represented as 4 in Fig. 2.8a and shown in detail in Fig. 2.8b). It comprises two portions: the upper joint and the lower joint. The upper joint can stretch out of the lower joint when the orbital radius increases. The upper motor (represented as 1 in Fig. 2.8a) drives the actuating shaft (represented as 3 in Fig. 2.8a), which is bounded together with the upper joint. The lower joint then drives the rotating spindle. The tail end of the spindle is a gas cylinder (represented as 5 in Fig. 2.8a). The press implemented on the polishing tool can be exactly controlled by the air pressure.

Fixed abrasive fabrication technology has been developed for tens of years, and was recently used in the polishing of optical component. Compared with loose abrasive, fixed abrasive polishing brings in many improvements in removal rate and draws wide public attention. The influence of the polishing parameters to material removal rate and surface roughness of fixed abrasive polishing has been studied extensively. Based on CCOS technology, people take the place of pitch pads with fixed abrasive diamond pellets (FADPs), which have been developed into fixed abrasive polishing. Cylinder pellets are bonded uniformly with a quantity of diamond abrasive, cemedin, and additive according to a specific ratio. The parameters of diamond pellets we used are listed in Table 2.2. The diamond granularity

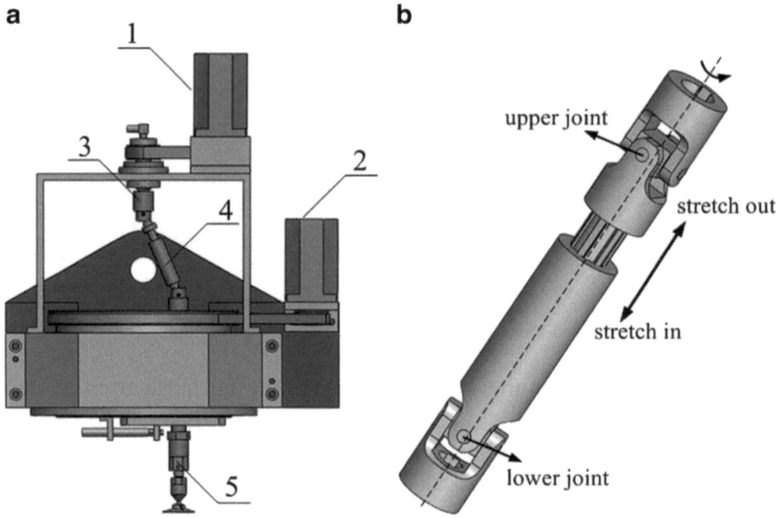

Fig. 2.8 Structure diagrams of (**a**) the fabrication unit; (**b**) the retractable universal joint

Table 2.2 Parameter list of fixed abrasive diamond pellets

Pellet kind	Diamond granularity (μm)	Bonding agent	Pellet diameter (mm)	Concentration (%)
W1.5	1.5	Resin	6	10
W3	3	Resin	6	10
W5	5	Resin	6	10
W17	17	Metal	6	10

varies from 1.5 to 17 μm, and all the concentrations of abrasive are 10 %. Resin and metal are adopted as cemedin. Continuously supplied deionized water is adopted as a coolant in all experiments. This structure characteristic brings in two advantages for TIFs.

1. *High stability*: Abrasive particles are uniformly distributed in the pellets, not just bonded in the outside surface. The thickness of pellets is 4 mm, and could be used for hundreds of hours. The external abrasives of pellets engender material removal by brittle fracture. When the external abrasives are worn down, sub-surface abrasives reach the top and keep removing material. This kind of removal feature guarantees the stable concentration distribution of abrasives, and then brings in highly stable material removal.
2. *High removal rate*: The abrasives are fixed on a cylinder rather than in the dissociative status, which could provide much larger shear force. Simultaneously, the ubiquitous splash phenomenon of abrasive slurry in loose abrasive polishing is avoided, which makes fixed abrasive polishing allow high speed (it is easy to achieve 1,000 rpm). These features certainly bring in a higher removal rate than loose abrasive.

Table 2.3 Fixed abrasive polishing pads with different distribution models

Model number	Sketch map	Optimized *f*	Optimized *g*	Optimized 3D-TIFs
1	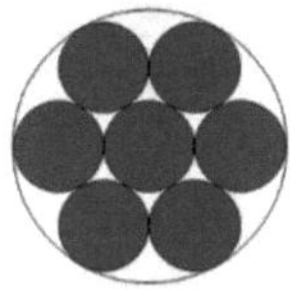	−3	0.8	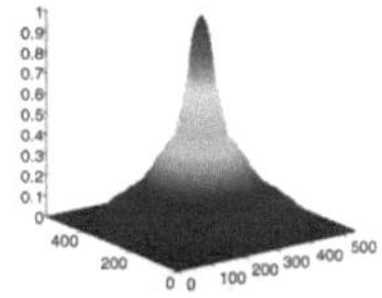
2		−3	0.8	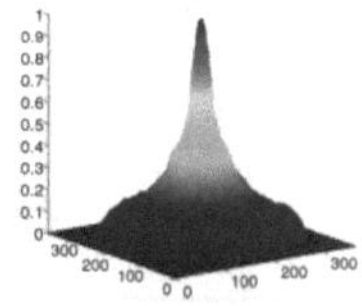
3	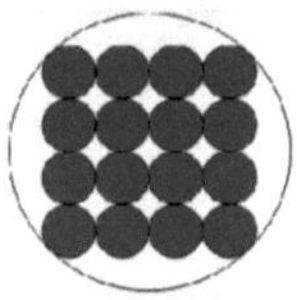	−3	0.7	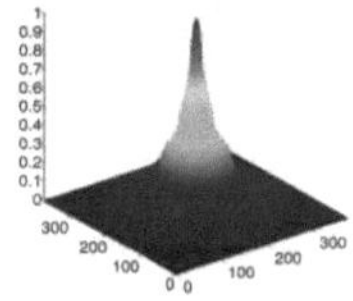
4	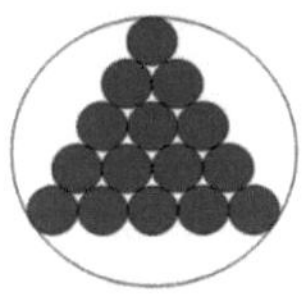	−5	0.7	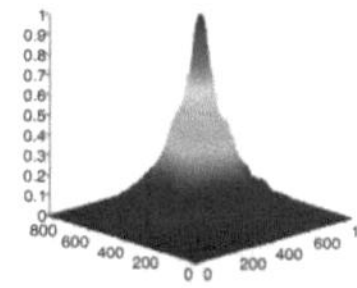
5	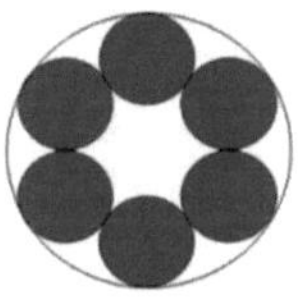	−5	0.6	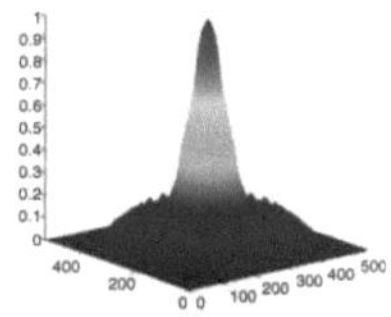
6	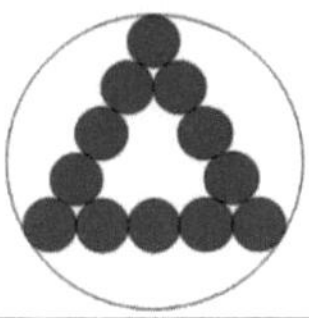	−4	0.6	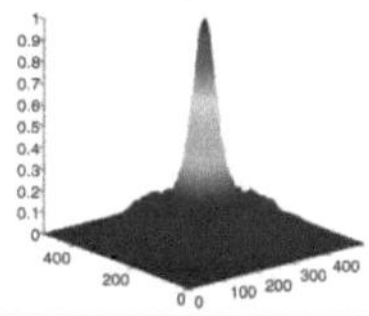

3D three-dimensional, *TIF* tool influence function

In the first step, the pellets are combined by hands into a polishing pad without a perfect flat or spherical surface, and could not be used for practical polishing. The next step, the pads should be modified to be a flat or spherical surface, by hand or machine, utilizing a flat or spherical steel plate to rub the pad rigidly. By this preprocess, the pad's surface could be coincident in curvature with SiC mirrors.

The polishing pads could be flexibly designed into different shapes, which provide us with different TIFs. To obtain the necessary TIF, we could make arbitrary changes in the amount of pellets, total aperture, position of the pellets and so on. According to the design experience of pitch pads, we design a mass of fixed abrasive polishing pads with different distribution models, such as circular, annular, square, triangle, and triangle annular, etc. These models are listed in Table 2.3. Some practical polishing pads are shown in Fig. 2.9.

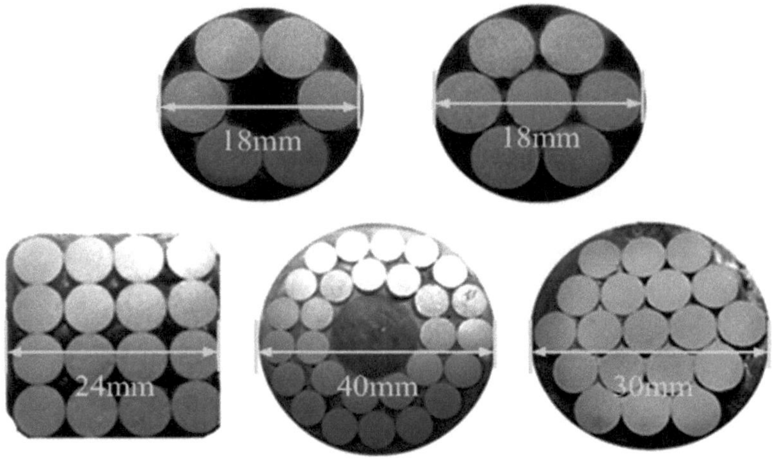

Fig. 2.9 Some practical polishing pads

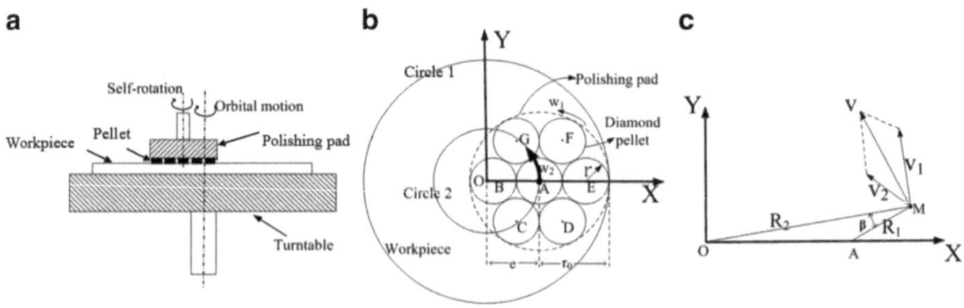

Fig. 2.10 The TIF model construction for fixed abrasive diamond pellet tool. *TIF* tool influence function

For fixed abrasive micro-grinding and polishing, the TIF model can be constructed as follows. As shown in Fig. 2.10, a calculation algorithm under the planetary motion model is established as follows.

In this model, the polishing pad moves in self-rotation and orbital motion. *Circle 1* is the polishing area in one orbital period, and *Circle 2* is the orbital trajectory. In the Cartesian coordinate *XOY*, point *O* is the orbital center. Points *A*, *B*, *C*, *D*, *E*, *F* and *G* represent every pellet's geometrical center, respectively—we call these center points of pellets (CPP). Obviously, point *A* also plays the role of the self-rotation center. The self-rotation angular speed of the polishing pad is w_1, and orbital angular speed is w_2. The radius of the single pellet is *r*, and the radius of the polishing pad is $r_0 = 3 \cdot r$. The acentric distance is *e*. In Fig. 2.10c, point *M* is an arbitrary point in the polishing area. The distance between *M* and *O* is R_2, and between *M* and *A* is R_1. v_1, v_2 represent self-rotation and orbital speed of *M*, respectively.

Two parameters are defined which are significant for the characteristics of TIFs. The first is the ratio between self-rotation and orbital annular speed, expressed as shown in Eq. (2.10). The other is the eccentricity, which could be expressed as the ratio between acentric distance and the pad's radius, which is presented in Eq. (2.11).

$$f = \frac{w_1}{w_2} \tag{2.10}$$

$$g = \frac{e}{r_0} \tag{2.11}$$

The orbital period is $T = 2\pi/w_2$. Divide the period T into n portions, one single portion $\Delta t = T/n$. Assume the initial position is the position as shown in Fig. 2.10, the TIF model construction. During the orbital period T, in a certain time of t, where $t = i \cdot \Delta t (i = 1, 2, 3 \ldots, n)$, the coordinates of CPPs during one period could be expressed as Eq. (2.12).

$$\begin{cases} \begin{cases} X_A = e - e \cdot \left(1 - \cos\left(w_2 \cdot t\right)\right) \\ Y_A = e \cdot \sin\left(w_2 \cdot t\right) \end{cases} \\ \begin{cases} X_B = e - 2 \cdot r - e \cdot \left(1 - \cos\left(w_2 \cdot t\right)\right) + 2 \cdot r \cdot \left(1 - \cos\left(w_1 \cdot t\right)\right) \\ Y_B = 0 + e \cdot \sin\left(w_2 \cdot t\right) - 2 \cdot r \cdot \left(\sin\left(w_1 \cdot t\right)\right) \end{cases} \\ \begin{cases} X_C = e - r - e \cdot \left(1 - \cos\left(w_2 \cdot t\right)\right) + r - 2r \cdot \sin\left(\pi/6 - w_1 \cdot t\right) \\ Y_C = -3^{1/2} \cdot r + e \cdot \sin\left(w_2 \cdot t\right) - \left(2r \cdot \cos\left(\pi/6 - w_1 \cdot t\right) - 3^{1/2} \cdot r\right) \end{cases} \\ \begin{cases} X_D = e + r - e \cdot \left(1 - \cos\left(w_2 \cdot t\right)\right) + 2r \cdot \cos\left(\pi/3 - w_1 \cdot t\right) - r \\ Y_D = -3^{1/2} \cdot r + e \cdot \sin\left(w_2 \cdot t\right) + \left(3^{1/2} \cdot r - 2r \cdot \sin\left(\pi/3 - w_1 \cdot t\right)\right) \end{cases} \\ \begin{cases} X_E = e + 2 \cdot r - e \cdot \left(1 - \cos\left(w_2 \cdot t\right)\right) - 2 \cdot r \cdot \left(1 - \cos\left(w_1 \cdot t\right)\right) \\ Y_E = 0 + e \cdot \sin\left(w_2 \cdot t\right) + 2 \cdot r \cdot \left(\sin\left(w_1 \cdot t\right)\right) \end{cases} \\ \begin{cases} X_F = e + r - e \cdot \left(1 - \cos\left(w_2 \cdot t\right)\right) + 2r \cdot \sin\left(\pi/6 - w_1 \cdot t\right) - r \\ Y_F = 3^{1/2} \cdot r + e \cdot \sin\left(w_2 \cdot t\right) + \left(2r \cdot \cos\left(\pi/6 - w_1 \cdot t\right) - 3^{1/2} \cdot r\right) \end{cases} \\ \begin{cases} X_G = e - r - e \cdot \left(1 - \cos\left(w_2 \cdot t\right)\right) + r - 2r \cdot \cos\left(\pi/3 - w_1 \cdot t\right) \\ Y_G = 3^{1/2} \cdot r + e \cdot \sin\left(w_2 \cdot t\right) - \left(3^{1/2} \cdot r - 2r \cdot \sin\left(\pi/3 - w_1 \cdot t\right)\right) \end{cases} \end{cases} \tag{2.12}$$

We can obtain the motion trajectory of every CPP through Eq. (2.12). During T, assume the total contiguous time between M and the polishing pad is t_M, initialize $t_M = 0$. In time t, we can easily obtain the length of line segments $\overline{MA}$, $\overline{MB}$, $\overline{MC}$, $\overline{MD}$, $\overline{ME}$, $\overline{MF}$, $\overline{MG}$, if they satisfy Eq. (2.13), then M and the polishing pad are contiguous in time t; in other words, t is a valid time segment, then $t_M = t_M + t$. Circulate every t to get t_M. The intersection angle between O and the circular arc path of M in the polishing pad is $2\alpha = 2 \cdot t_M/n \cdot \pi$. Relative speed between O and M could be expressed as Eq. (2.14).

$$\overline{MA} < r \middle| \overline{MB} < r \middle| \overline{MC} < r \middle| \overline{MD} < r \middle| \overline{ME} < r \middle| \overline{MF} < r \middle| \overline{MG} < r \tag{2.13}$$

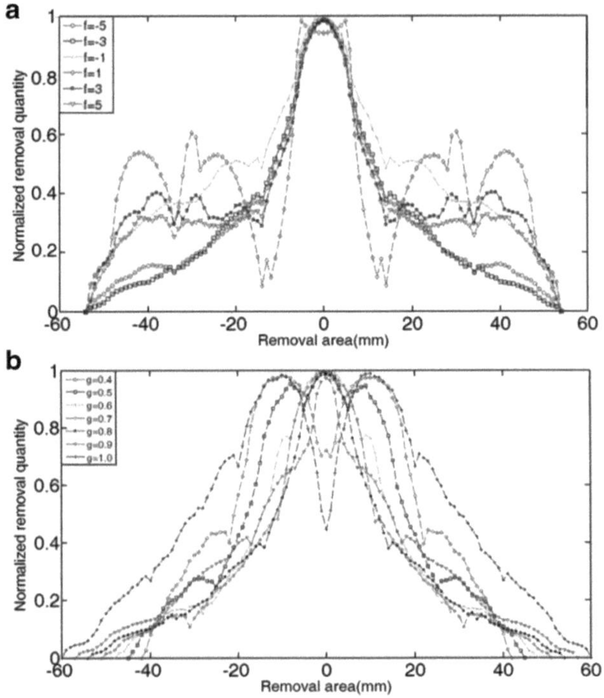

Fig. 2.11 (**a**) Different removal functions (RFs) under different speed ratios; (**b**) different RFs under different eccentricity ratios

$$
\begin{aligned}
|\boldsymbol{v}| &= |\boldsymbol{v}_1 + \boldsymbol{v}_2| \\
&= \left(v_1^2 + v_2^2 - 2v_1v_2\cos\beta\right)^{1/2} \\
&= \left((R_1w_1)^2 + (R_2w_2)^2 - 2R_1w_1R_2w_2\cos\beta\right)^{1/2}
\end{aligned} \tag{2.14}
$$

Where: $\beta = \arccos\left(\frac{R_1^2+R_2^2-e^2}{2R_1R_2}\right)$, $R_2 = (R_1^2 + e^2 - 2R_1e\cos\alpha)^{1/2}$.

The contiguous time of M is expressed as Eq. (2.15).

$$
\mathrm{d}t = \mathrm{d}\alpha/w_2 \tag{2.15}
$$

Combine Eqs. (2.14) and (2.15), and the Preston function, the TIF can be obtained by Eq. (2.16).

$$
\mathrm{TIF}(r) = \mathrm{d}z = \int_{-\alpha}^{\alpha} K \cdot P \cdot v/w_2 \cdot \mathrm{d}\alpha \tag{2.16}
$$

where TIF(r) is the tool influence function, K is the Preston constant, P is the pressure between the polishing pad and workpiece, and v is the relative speed between the polishing pad and workpiece in M.

The theoretical TIFs under different parameters are shown in Fig. 2.11. Figure 2.11a represents different TIFs under different speed ratios while the

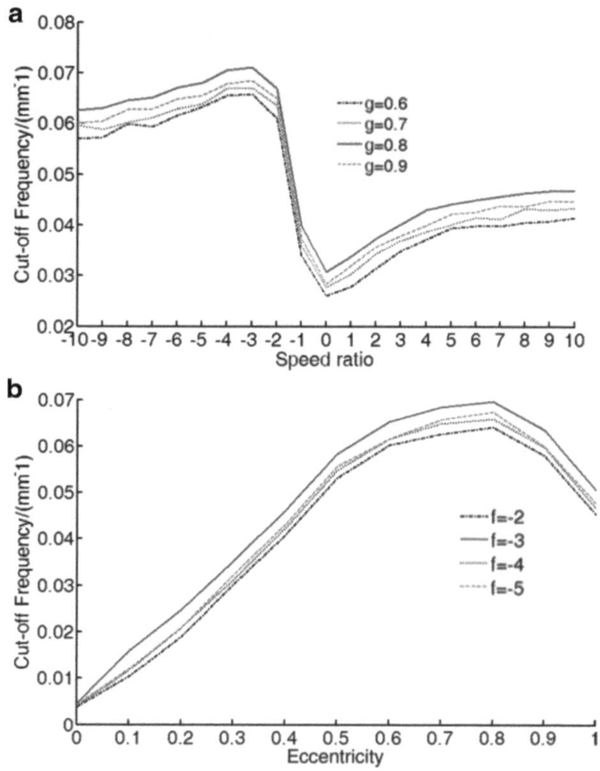

Fig. 2.12 Cutoff frequency about *f* and *g*

eccentricity ratio $g = 0.8$. Because of the desire for higher center removal rate, the optimized speed ratio is $f = -3$ or $f = -5$. Figure 2.11b shows different TIFs under different eccentricity ratios while speed ratio $f = -3$, we can find that $g = 0.7$ and $g = 0.8$ may be optimized selections.

2.3.1.2 Analysis and Optimization

Take model 1 in Table 2.3 for example, the cutoff frequency about f and g is calculated and plotted, respectively, in Fig. 2.12. Figure 2.12a indicates the cutoff frequency curves of $g = 0.6$, $g = 0.7$, $g = 0.8$, $g = 0.9$, while the speed ratio varies from -10 to 10 by 1. When the speed ratio is negative, the cutoff frequency shows much higher than positive speed ratio. For higher cutoff frequency, $f = -3$ could be taken as the best speed ratio from these curves. Figure 2.12b reveals the cutoff frequency curves of $f = -2, f = -3, f = -4, f = -5$, while eccentricity varies from 0 to 1.0 by 0.1. Here, we set the scale of all the TIFs as 40 mm. The maximum cutoff frequency appears in $g = 0.8$.

Therefore, we generally take $f = -3$ and $g = 0.8$ as the optimized parameters. They are generally coincident to the prediction in the above analysis.

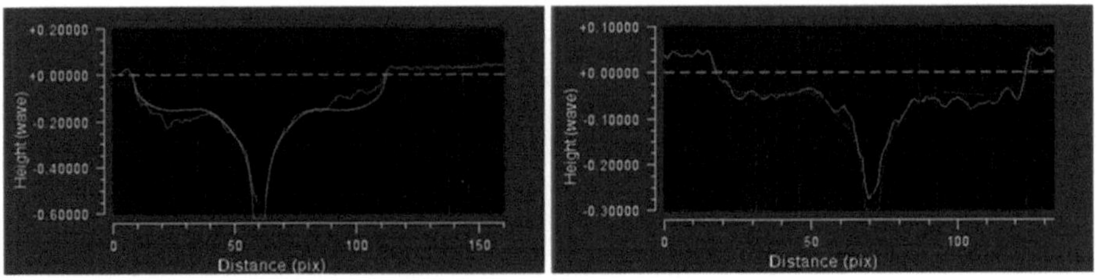

Fig. 2.13 Contrast of practical removal profile and emulational profile

2.3.1.3 Experimental Validation

Experiments are conducted to validate the effectiveness of the model on forecasting the removal profile of small-tool polishing. Detailed parameters are concluded as follows: the optical segment is a 200 mm SiC, which was per-polished to be a fine plane surface. The experiments adopted 10 weight percent (wt%) CeO_2, press kept 30 N. Tool size was 30 mm, and the polishing material was 4# pitch. The eccentric distance was 12 mm, and the orbital and spinning velocity were 60 rpm and 480 rpm, respectively. The experimental result was measured by Zygo GPI and shown in Fig. 2.13 (*left*). We find that the profile of TIF is highly identical to the emulational profile. The second polishing process decreased the spinning velocity to 300 rpm, and obtained results as shown in Fig. 2.13 (*right*), which also has high similarity.

We performed a series of experiments on reaction-bonded SiC mirrors to validate the algorithm and optimized parameters; results of TIFs of model 1 and 5 are presented in Fig. 2.14. Corresponding experimental parameters are listed in Table 2.4.

As shown in Fig. 2.14, the TIFs with a high removal center are obtained, which could help raise the convergence rate in a polishing run. The coincidence of profile between experimental and theoretical TIFs is generally satisfactory. The small difference is mainly caused by the unsuitability between the optical surfaces and polishing pads, which inevitably exists. These processes validate the effectiveness of the proposed algorithm.

A polishing pad of model 5 with pellets of diamond granularity 3 μm is used to validate the stability of TIFs as usage time increases. The experimental parameters are listed in Table 2.5 and the experimental results are shown in Fig. 2.15.

In Fig. 2.15a, the upper removal shape (shown as I) is experiment 1, in which the usage time of polishing pads is $t = 0 - 10$ min. The lower removal shape (shown as II) is experiment 2, whose usage time is $t = 300 - 310$ min. During the usage time from 10 to 300 min, the pad is used to polish another SiC workpiece. In Fig. 2.15b, the left curve is a 2D profile of experiment 1, and the right curve is a 2D profile of experiment 2. We could find the result shows high stability both in profile shape and peak removal magnitude as tool usage time increases. This means, during

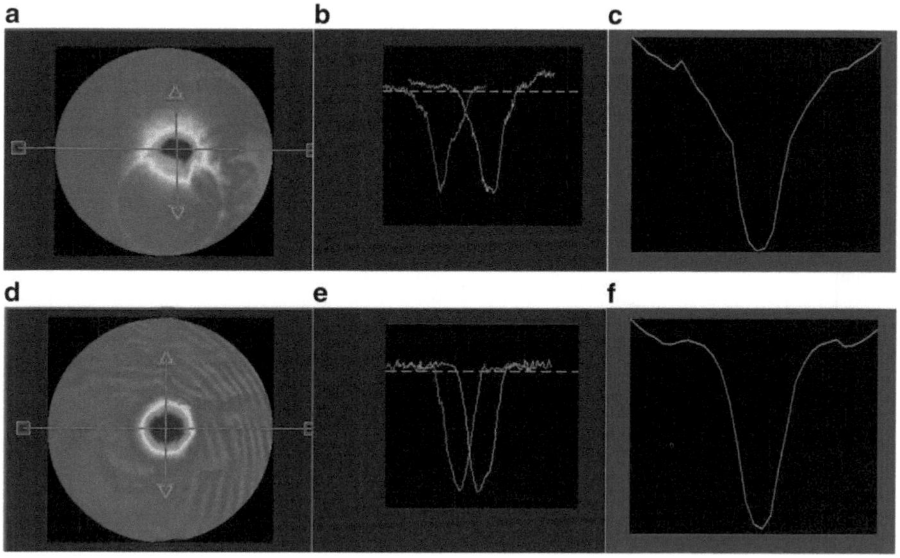

Fig. 2.14 Results of TIFs of models 1 and 5. *TIFs* tool influence functions

Table 2.4 Corresponding experimental parameter of models 1 and 5

Number	Model	Pellets	f	g	Pressure (N)	Time (min)
1	1	W3	−3	0.8	30	10
2	5	W3	−3	0.8	30	10

Table 2.5 Experimental parameters for tool influence function stability

Remarks	Usage time (min)	f	g	Pressure (N)
Experiment 1	0–10	6	0.8	20
Experiment 2	300–310	6	0.8	20

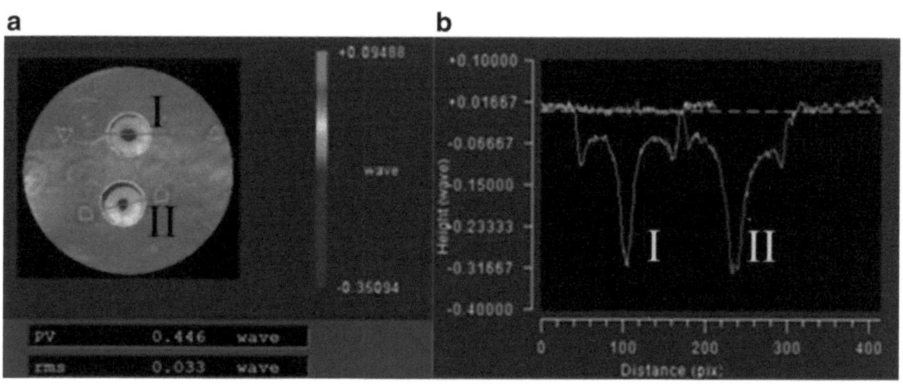

Fig. 2.15 (**a**) *Top* view removal shape; (**b**) *left* curve is a 2D profile of experiment 1, and the *right* curve is a 2D profile of experiment 2. *2D* two-dimensional

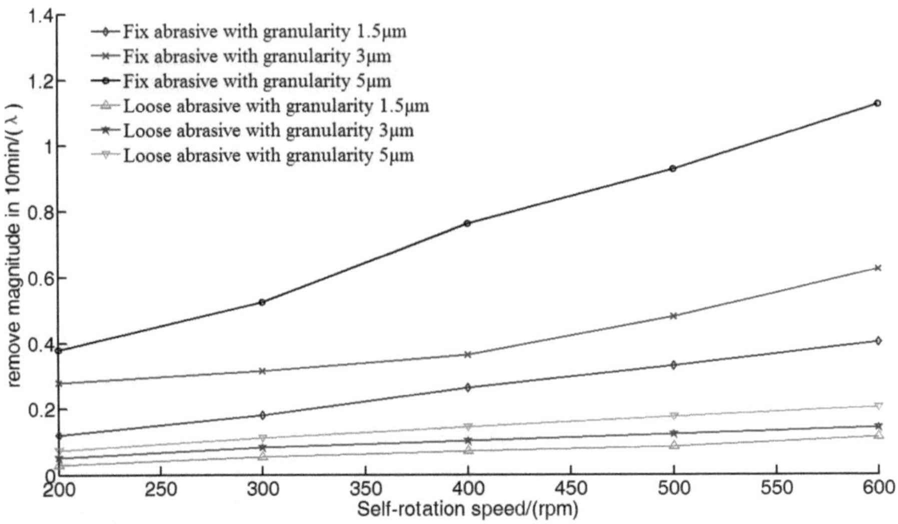

Fig. 2.16 Results of ten experiments

Table 2.6 Conditions of experiments for removal rate

Number	Polishing pads	Granularity (μm)	Self-rotation speed (rpm)	*f*	*g*	*F*(*N*)
1	Diamond pellets	1.5	200–600	−3	0.8	30
2	Diamond pellets	3	200–600	−3	0.8	30
3	Diamond pellets	5	200–600	−3	0.8	30
4	Pitch + diamond abrasive	1.5	200–600	−3	0.8	30
5	Pitch + diamond abrasive	3	200–600	−3	0.8	30
6	Pitch + diamond abrasive	5	200–600	−3	0.8	30

rpm revolutions per minute

a relatively long polishing time (5 h or more), the fixed abrasive could provide highly stable TIFs for the polishing of SiC mirrors.

Tens of experiments are conducted to compare the removal rate of loose abrasive and fixed abrasive polishing. The results are plotted in Fig. 2.16. Table 2.6 lists the conditions of experiments. Experiments 1–3 are fixed abrasive polishing with granularity of 1.5, 3, 5 μm, respectively, and experiments 4–6 use pitch pad and diamond loose abrasives of 1.5, 3, 5 μm, respectively. In Fig. 2.16, every experiment includes five portions, whose self-rotation speed varies from 200 to 600 rpm. As an example, when the self-rotation speed is 500 rpm, and with 5 μm granularity, the loose abrasive removal magnitude in 10 min is 0.176 λ, and fixed abrasive is 0.926 λ, which could provide more than five times the removal rate. For other experimental parameters, fixed abrasive could generally achieve a removal rate

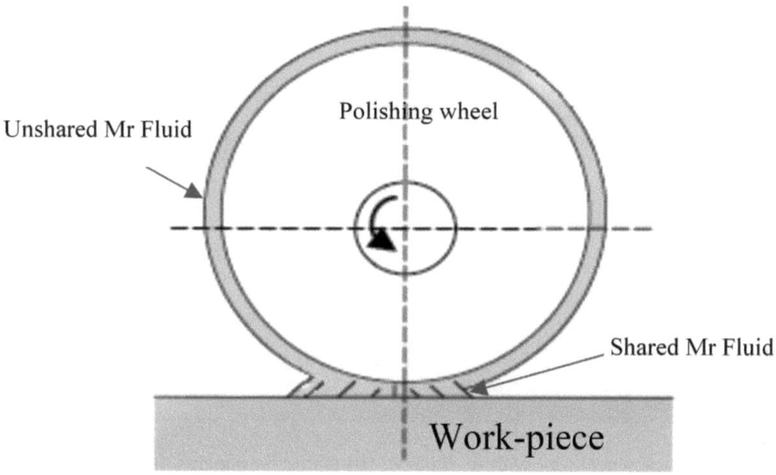

Fig. 2.17 Basic principle of MR fluid-assistive polishing. *MR* magnetorheological

about five times more than the loose abrasive, which could help save a great deal of fabrication time.

2.3.2 *Magnetorheological Finishing System*

2.3.2.1 Model Construction

Magnetorheological (MR) fluids are composed of small magnetic particles in disperse phase suspended in non-conducting carrier liquids. Under the effect of magnetic field, the magnetic particles are polarized and organized into chains or ribbons of particles. As such, the structure and rheological properties of MR fluids, such as viscosity and stress, can be dramatically altered. It normally takes only milliseconds for MR fluid passing through a magnetic field to stiffen up, and then return to its original state upon exiting, again in milliseconds. The phenomenon can thus be utilized for sub-aperture polishing of optical surfaces with the mixing of ultra-fine abrasive (non-magnetic) particles to the fluid. In this regard, Cheng et al. [4, 5] designed a wheel-tool for MR fluid-assistive polishing. As shown in Fig. 2.17, the portion of magnetically stiffened fluids becomes a polishing tool at the region of closest separation between the tool and the workpiece, where the fluid ribbon is squeezed from about 2 mm to 1 mm in thickness. As generated, the resulting sub-aperture tool has the advantage of producing significant shear stress and polishing pressure while automatically conforming closely to the local curvature being polished.

One precondition for successful MR polishing is the generation of suitable gradient of magnetic field by the wheel tool (Fig. 2.17). Researchers have proposed

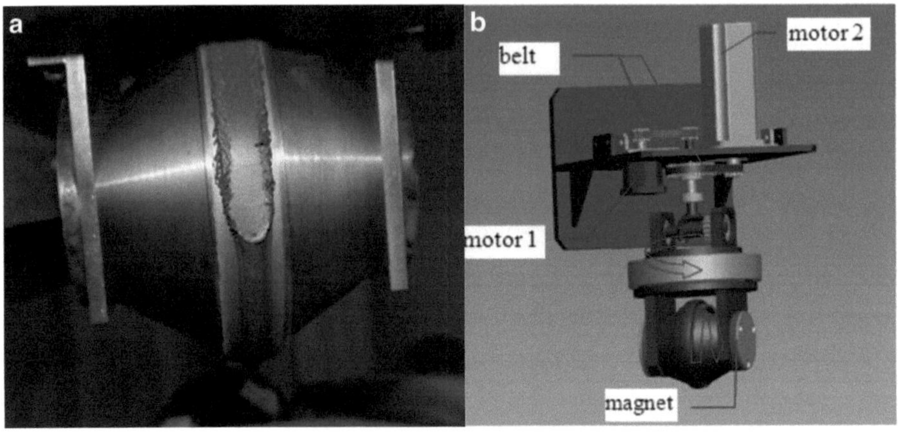

Fig. 2.18 (**a**) Actual wheel tool with workpiece; (**b**) main components of the tool design

various designs, e.g., Golini [6] and Cheng et al. [7], and conducted experiments validating their polishing performance. The previous works, however, all adopted tools that rotate about only one axis (the self-rotating axis). Their emphases were focused on testing the MR processes with different materials and studying the resulting performance. Little work was reported on using other wheel-tool design and understanding the effects of process parameters on MR polishing, analytically or experimentally.

The tool we designed adopts a two-axis wheel-shaped polishing tool, as shown in Fig. 2.18, for analysis and experimentation. Figure 2.18a shows the picture of the actual hardware and Fig. 2.18b depicts the main components of the setup. During MR polishing, the wheel tool is driven by motor 1 to rotate (with angular speed w_2) about its self-rotating axis through a belt and bevel gear setup, and at the same time by motor 2 to rotate (with angular speed w_1) about the co-rotating axis. Figure 2.19 shows the internal structure underneath the tool cover and the magnetic fields from the tool design. Referring to Fig. 2.19a, B1–B2 is a pair of ring magnets arranged symmetrically on the two sides of the wheel tool. The orientation is such that the tool's self-rotating axis is normal to the ring magnets at their centers. A1–A2 is a pair of rectangular magnets located near the bottom of the tool cover at close distance to the disc inside. The disc is an 'I' shaped wheel made of non-conducting material to carry MR fluids, and rotates about the self-rotating axis. The design has A1–A2 producing stronger magnetic intensity than B1–B2. The magnetic field lines around the disc can be determined from theory, as shown in the blue dashed lines in Fig. 2.19a. The two magnetic fields from A1–A2 and B1–B2 overlap at the bottom of the disc, resulting in a localized magnetic field with stronger intensity on top of a weaker field that is more broadly distributed over the disc surface. At the top apex of the disc, in particular, the field has minimum intensity. This physical description is verified by Fig. 2.19b, which shows the magnetic field lines simulated using finite element analysis. Note that for enhanced

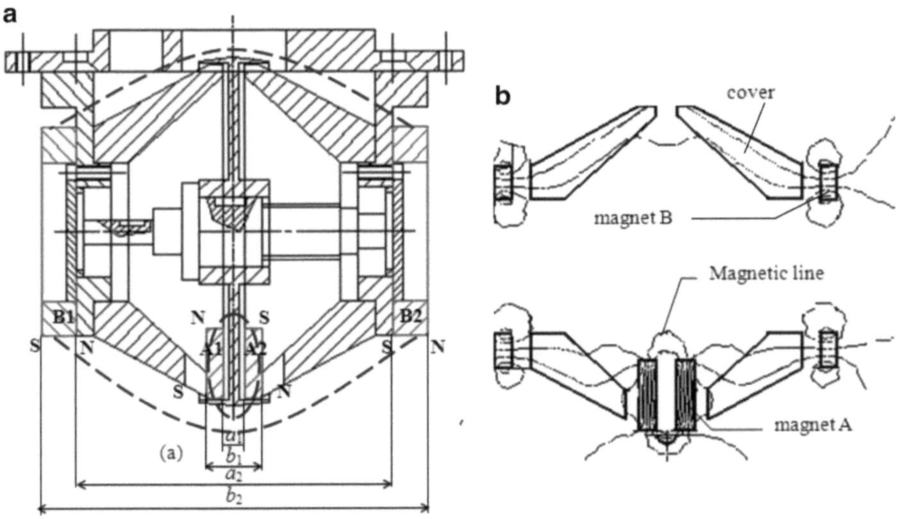

Fig. 2.19 (**a**) Internal structure and magnet distribution of the wheel tool; (**b**) magnetic field lines generated by finite element analysis (FEA) methods

polishing, the overall magnetic field should transverse the working gap between the polishing wheel and the workpiece in a direction as perpendicular to the workpiece surface as possible.

During actual polishing, the workpiece would sit on top of a turntable rotating at a relatively slow speed (to w_1 and w_2). MR fluid ribbon adhering to the surface of the wheel disc is transported into the gap between the disc surface and the workpiece. Under the action of the double magnetic fields, the stiffened fluid region forms a transient polishing zone or spot on the workpiece surface. Surface material within this polishing spot is removed by the self-rotating motion. The role of the co-rotating motion is to continually re-orient the direction of the self-rotation motion to smooth out any angular preference in the polishing spot. Moreover, once the fluid ribbon is transported outside the gap by the disc's self-rotation, it tends to disband and lose the ability to carry the abrasive particles because of reduction in the magnetic field intensity. The role of the magnetic field B1–B2, although at weaker intensity, is to maintain the shape of the ribbon to a certain extent, and to help with the loosening, recombination, and replenishing of the abrasive particles.

Polishing is a complex process and both relative velocity and pressure between the tool and the workpiece are important factors to consider. As the tool travels over the workpiece surface, the material can be obtained from the convolution of the material removal function with the dwell time along the polishing path. The material removal function is given by the Preston function discussed in Chap. 1.

The wheel tool's relative velocity is a combination of its self-rotation and co-rotation motions. The relatively slower self-rotation of the turntable on which the workpiece is sitting is omitted. Figure 2.20 depicts the transient polishing spot

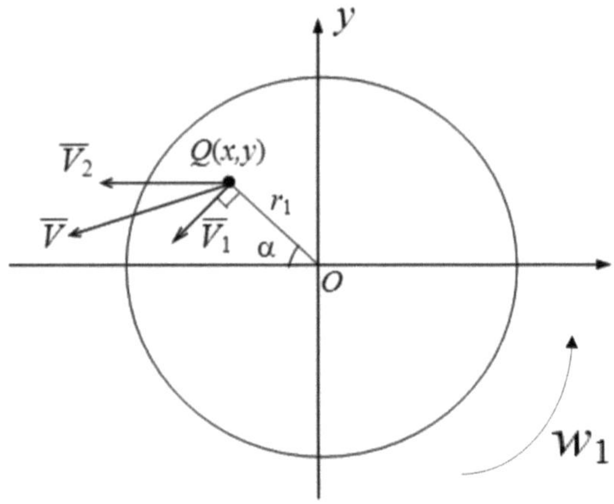

Fig. 2.20 Polishing spot on workpiece indicating relative velocity due to wheel tool rotations

mentioned above on the workpiece surface. The relative velocity at an arbitrary point $Q(x,y)$ inside the polishing spot can be expressed as

$$\vec{V} = \begin{cases} \vec{V_2} + \vec{V_1} \\ w_1 r_1 \rightarrow \vec{V_1} \\ w_2 r \rightarrow \vec{V_2} \end{cases} \tag{2.17}$$

where $\overline{V}_1$ and $\overline{V}_2$ are the linear velocities due to the co-rotation and self-rotation motion, respectively, at point Q, r_1 is the distance from point Q to the center O of the polishing spot, and r is the wheel-disc's semi-diameter. Note that O here is also the center of the co-rotation motion—the co-rotation axis would be the normal to the polishing spot at this point. Thus,

$$V = \sqrt{(w_1 r_1)^2 + (w_2 r)^2 - 2 w_1 w_2 r_1 r \cos\left(\frac{\pi}{2} + \alpha\right)} \tag{2.18}$$

The normal pressure P inside the polishing spot is a very complicated quantity. It can be divided into two parts: the hydrodynamic pressure P_h and the magnetic pressure P_m, i.e.,

$$P = (P_h + P_m) \cdot \frac{\sqrt{r^2 - x^2}}{r} \tag{2.19}$$

The multiplicative factor in Eq. (2.19) serves to convert the pressure exerted along the radial direction of the disc to the normal direction of the polishing spot. As the MR fluid is incompressible, its contribution to P_h can be ignored. P_h can hence be simplified, as shown by Chi et al. [8]:

$$P_h = \frac{2\eta V}{\left(d + \frac{x^2}{2r}\right)^2} \tag{2.20}$$

where d is the minimum gap between the bottom of the wheel disc and the surface of the workpiece, and η is the working viscosity of the MR fluid during the process.

The magnetic fields generated by the wheel tool in Fig. 2.19 can be derived roughly from the theory. Approximating the magnet arrangement for both A1–A2 and B1–B2 as parallel plates, the distribution of the magnetic intensity H in-between A1–A2 and B1–B2 can be expressed as,

$$\begin{cases} H = \sum\limits_{n=1}^{\infty} - A_n \cos(\beta_n x) e^{-\beta_n y i} + \sum\limits_{n=1}^{\infty} A_n \sin(\beta_n x) e^{-\beta_n y j} \\ A_n = K_n \beta_n = \dfrac{2B_g \sin\left(\beta_n \dfrac{a}{2}\right)}{\pi(2n-1)\mu_0} \\ \beta_n = \dfrac{2n-1}{b}\pi \\ K_n = \dfrac{2b \cdot B_g \cdot \sin\left(\beta_n \dfrac{a}{2}\right)}{\pi^2 \cdot (2n-1)^2 \cdot \mu_0}, \quad n = 1, 2, 3 \ldots \end{cases} \tag{2.21}$$

where n is the iterative order number, A_n, β_n, and K_n are the corresponding coefficients, B_g is magnetic induction intensity, μ_0 is the permeability of vacuum, and a is the inner distance and b the outer distance between the two magnets in A1–A2 or B1–B2, as the case may be. Consider then the interaction force between aggregated magnetic particles under the action of the magnetic field. A magnetic particle, e.g. a spherical carbonyl iron (CI) particle, of radius R and magnetic permeability μ_p will have a magnetic moment of

$$m = 4\pi\mu_0\mu_f R^3 \frac{\mu_p - \mu_f}{\mu_p + 2\mu_f} H \tag{2.22}$$

where μ_f is the magnetic permeability of the carrier liquid. Let ϕ be the fractional volume of magnetic particles in the MR fluid, the magnetic pressure P_m is hence, approximately,

$$P_m = \phi \int_0^H \frac{m}{\frac{4}{3}\pi R^3} \mathrm{d}H = 3\phi\mu_0\mu_f \frac{\mu_p - \mu_f}{\mu_p + 2\mu_f} \int_0^H H \mathrm{d}H \tag{2.23}$$

Table 2.7 Design and operating parameters of the two-axis wheel tool

Parameter	Value
Radius of carbonyl iron particles R (μm)	1–5
Volume fraction of carbonyl iron particles in fluid ϕ (%)	20–40
Volume fraction of CeO_2 particles in fluid ϕ (%)	3–8
Volume fraction of water in fluid ϕ (%)	50–75
Magnetic induction intensity of A1–A2 B_{g1} (T)	0.99
Magnetic induction intensity of B1–B2 B_{g2} (T)	0.35
Permeability of vacuum μ_0 (H/m)	$4\pi \times 10^{-7}$
Magnetic permeability of host fluid μ_f (H/m)	$4\pi \times 10^{-7}$
Magnetic permeability of magnetic particles μ_p (H/m)	2×10^{-3}
Distance between two insides of A1–A2 a_1 (mm)	6
Distance between two outsides of A1–A2 b_1 (mm)	26
Distance between two insides of B1–B2 a_2 (mm)	120
Distance between two outsides of B1–B2 b_2 (mm)	140
Wheel-disc's semi-diameter r (mm)	43
Gap between disc's bottom and the workpiece d (mm)	1–2.5
Co-rotational speed of the tool w_1 (rpm)	100
Self-rotational speed of the tool w_2 (rpm)	0–200

For the tool design supporting dual magnetic fields, P_m is composed of P_{m1} and P_{m2} individually produced by the magnets A1–A2 and B1–B2, i.e.

$$P_m = P_{m1} + P_{m2} \tag{2.24}$$

Table 2.7 lists the design and operating parameters of the two-axis tool adopted for the present study. Based on the values given, hydrodynamic pressure and magnetic pressure are computed at varying distances from the center point O of the polishing spot. They are plotted in Fig. 2.21, assuming a typical polishing spot of 5 mm radius. The magnetic pressure is about six orders of magnitude smaller than those of hydrodynamic pressure. This points to a simplification of the relative pressure P as,

$$P \approx P_h \cdot \frac{\sqrt{r^2 - x^2}}{r} = \frac{2\eta V}{\left(d + \frac{x^2}{2r}\right)^2} \cdot \frac{\sqrt{r^2 - x^2}}{r} \tag{2.25}$$

Accordingly, material removal is directly related to such process parameters as the minimum gap d between the bottom of the wheel disc and the surface of the workpiece, the relative velocity V, and the working viscosity of the MR fluid η. The main effect of the applied magnetic field is not on P_m but on increasing the viscosity η of the polishing process. Actually, the working viscosity is affected by both the magnetic field intensity and fractional volumes of different components of the fluid. In this regard, the field intensity to make MR fluid stiff enough to support the polishing process has been investigated by Cheng et al. [4, 7]. The following efforts will be devoted to studying the fractional volume composition of MR fluid, ensuring that a sufficient level of magnetic intensity is being applied.

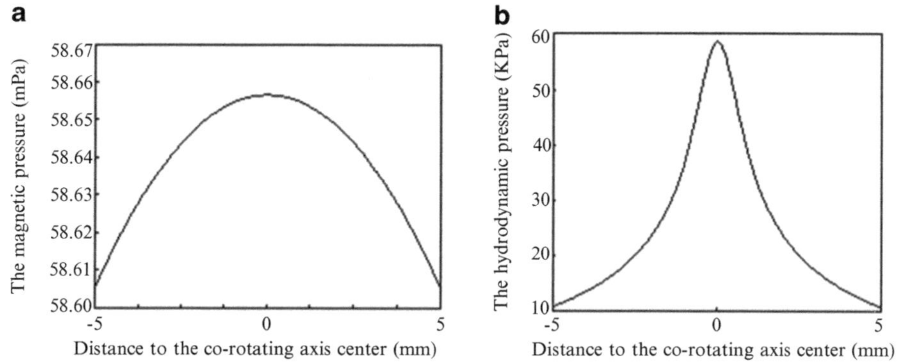

Fig. 2.21 (**a**) Plot of magnetic pressure and (**b**) hydrodynamic pressure as function of distance to the center of the polishing spot (Reproduced from Cheng et al. [5])

Table 2.8 Experiments to study material removal

Time (min)	w_2 (rpm)	d (mm)	ϕ of CI (%)	ϕ of CeO_2 (%)	ϕ of water (%)
10–60	150	1	33	6	57
30	0–200	1	33	6	57
30	150	1–2.5	33	6	57
30	150	1	20–40	6	57
30	150	1	33	3–8	57
30	150	1	33	6	50–75

rpm revolutions per minute, *CI* carbonyl iron

2.3.2.2 Experimental Validation

Experiments were conducted first to validate the general performance of the two-axis wheel tool and also to test out the best MR fluid composition for material removal. The experimental settings are listed in Table 2.8 and the results given in Fig. 2.22. A K9 glass parabolic lens of 60 mm diameter was used as the workpiece.

Figure 2.22a shows the increased amount of material being removed as a function of the polishing time. Figures 2.22b–f further show the maximum removal rate attained upon the varying of different design and process parameters. These plots show that maximum removal rate increases with the tool's self-rotational speed, the fractional volume of magnetic particle CI and abrasive particle CeO_2 in the MR fluid, and decreases with the increasing of the gap between the disc and the workpiece and the fractional volume of host water in MR fluid. For higher mixing ratio of abrasive particles in the fluid, more abrasives would be engaged in polishing within a unit time and hence a faster rate of material removal. On the other hand, when too much CI is mixed into the MR fluid, the fluidity would deteriorate and small clumps would accumulate under the action of the magnetic field, resulting in unsatisfactory polishing conditions. According to the experimental results, a tool

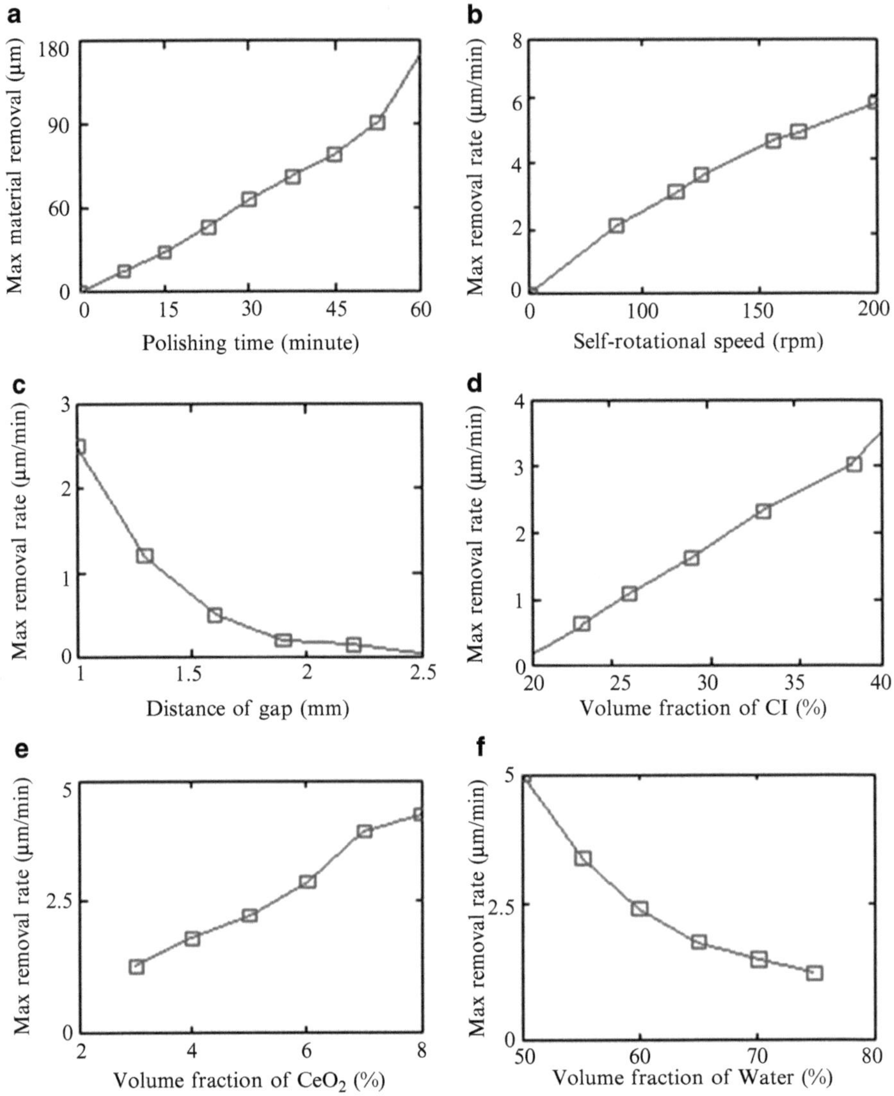

Fig. 2.22 Experimental results for the maximum material removal and the rate of versus (VS). (**a**) Polishing time; (**b**) the tool's self-rotational speed; (**c**) gap between the tool-disc's bottom and the workpiece; (**d**) volume fraction of CI; (**e**) volume fraction of CeO_2; (**f**) volume fraction of host fluid water in the MR fluid. *CI* carbonyl iron, *MR* magnetorheological (Reproduced from Cheng et al. [5])

speed ranging from 50 to 180 rpm, a gap distance ranging from 1 to 2 mm, and an MR fluid with 33 % CI, 57 % water, 6 % CeO_2, and 4 % stabilizing agent were deemed to provide the best removal performance.

Figure 2.23 shows the polishing performance on the K9 glass lens, taking into consideration the above findings. In Fig. 2.23a, it is easy to observe that the removal

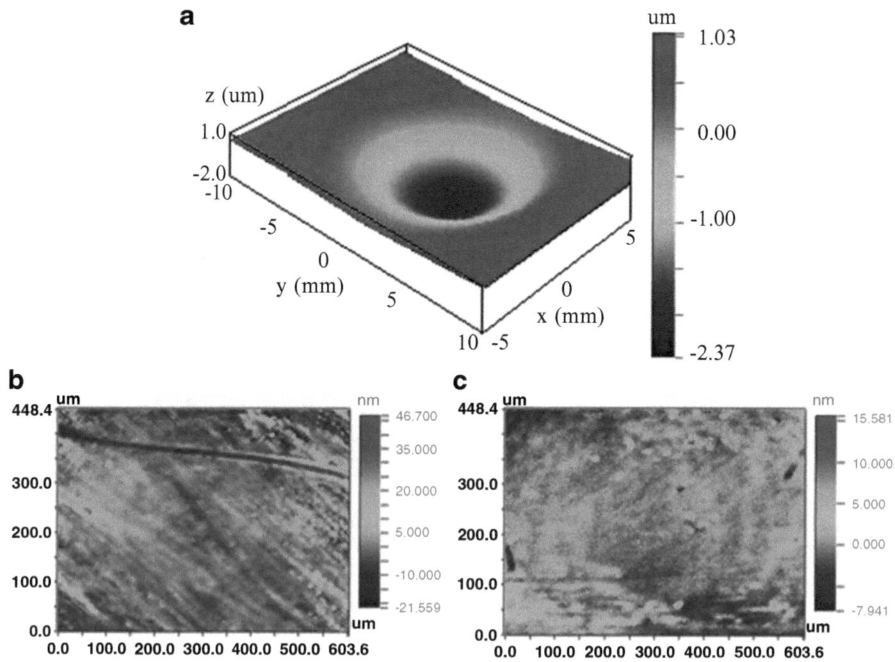

Fig. 2.23 3D profile of polishing spot (**a**) and surface roughness microstructure measured by a Wyko interferometer showing (**b**) initial Roughness average (Ra) = 3.8 nm, and (**c**) Roughness average (Ra) = 1.2 nm after 10 min of pre-polishing. *3D* three-dimensional (Reproduced from Cheng et al. [5])

function spot with a higher central peak would result in successful polishing. To achieve a smooth surface, we divide the process into pre-polishing and fine polishing stages. The pre-polishing uses a gap distance of 1 mm and a self-rotational speed of 180 rpm, whereas fine polishing resets the gap distance to between 1.5 and 2 mm while keeping the same self-rotation speed. It can be seen that surface roughness is reduced from Ra = 3.8 nm in Fig. 2.23b to Ra = 1.2 nm in Fig. 2.23c after 10 min of pre-polishing. The very apparent marks in Fig. 2.23b are readily smoothed out. The results confirm the overall effectiveness of the experimental process parameters and MR fluid composition in the quality polishing of optical surfaces using a two-axis wheel tool.

2.3.3 Magnetorheological Jet Polishing System

2.3.3.1 Model Constructions

The design of the jet tool with the electric current and steady jet is more complex in the eccentric rotation system than in the traditional system. Cheng et al. [9] reports a polishing tool with eccentric rotation motion, as shown in Fig. 2.24. This

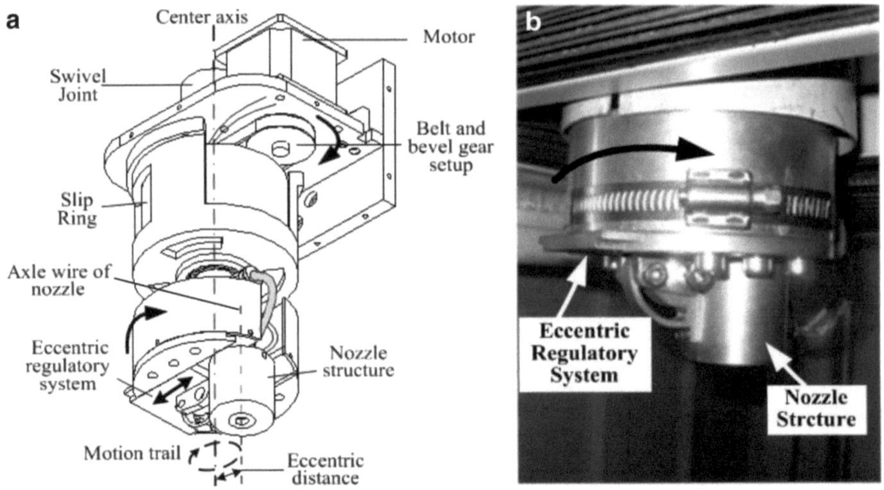

Fig. 2.24 Experimental setup. (**a**) computer aided design (CAD) of polishing setup; (**b**) main components of actual hardware

polishing tool can implement eccentric rotation motion mainly due to the utilization of eccentric regulatory system, swivel joint and slip ring. During the polishing process, the eccentric distance which depends on the diameter of the nozzle, is regulated by the eccentric regulatory system at first. This tool is then moved by high-precision machine to the target location on the glass for polishing. After that, eccentric regulatory system and nozzle structure will be driven to rotate around the center axis by the motor through belt and bevel gear setup. Meanwhile, the steady current will be transmitted from the external system to the coil via the slip ring, and the function of the swivel joint is similar to the slip ring, which transmits the continuous slurry from the external system to the nozzle (the coil and nozzle are in the interior of nozzle structure). The round jet is magnetized by the magnetic coil when it flows out of the nozzle. Therefore, the motion trail of the jet is circular, in particular, and its center is in the center axis as shown in Fig. 2.24.

According to the studies of Tricard et al. [10] and Kordonski et al. [11], the removal rate R at one point (x, y) can be expressed as below

$$R = kpv = k\frac{f}{\mu S}v = k\frac{D}{\mu} \tag{2.26}$$

where k is a constant parameter but depends on surface chemistry, workpiece material properties, and the slurry, etc., p is the normal pressure, v is the relative velocity between abrasives and the workpiece, f is the frictional force between the workpiece and the polishing lap, S is polishing zone, μ is the coefficient of friction, and D is the rate of work carried out at the surface.

When the MR jet passes beyond the magnetic field and is directed vertically towards the workpiece, the remanence declines fast and becomes very weak when

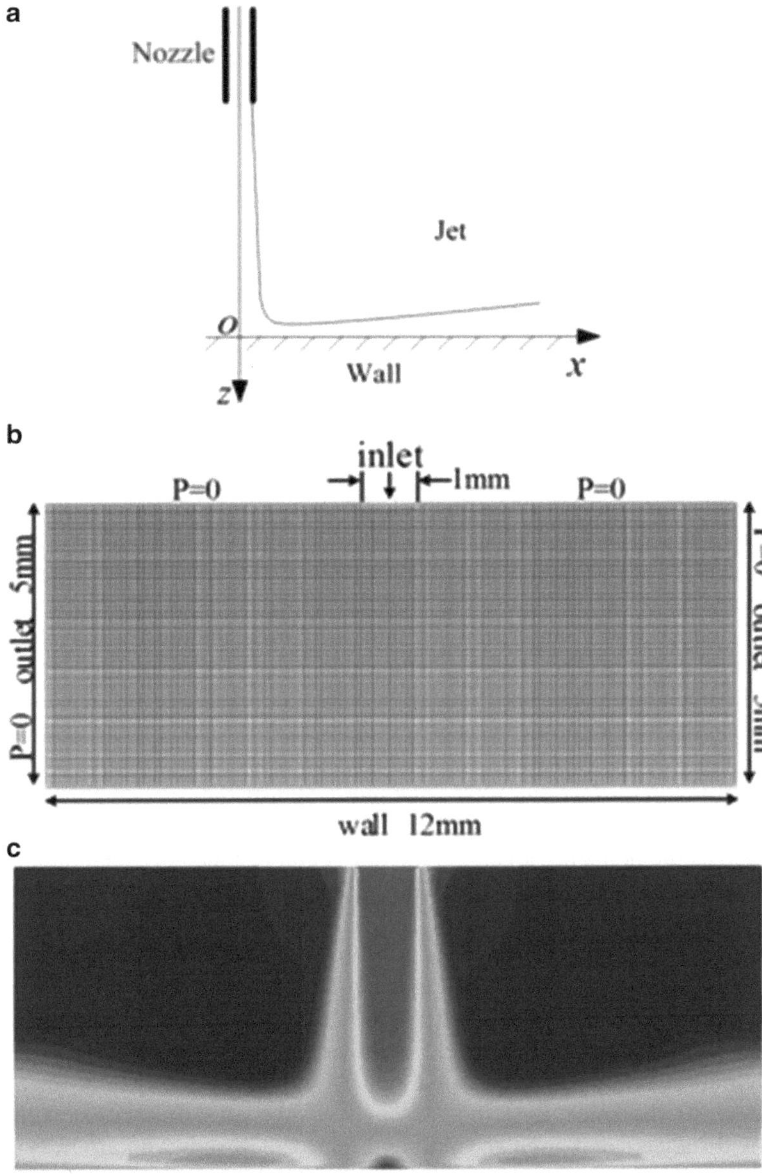

Fig. 2.25 Free impinging jet model. (**a**) Simple mode of jet; (**b**) mesh and boundary of the model; (**c**) simulation result with the jet velocity 20 m/s and nozzle diameter 1 mm

the jet impinges on the workpiece. Hence, the magnetic field can be ignored at the impingement area on the workpiece, and an assumption is made that the free impinging jet polishing theory (proposed by Becker et al. [12]) can be employed approximately nearby the workpiece where the jet remains coherent. A typical jet polishing model is shown in Fig. 2.25. The coherent jet hits the surface vertically.

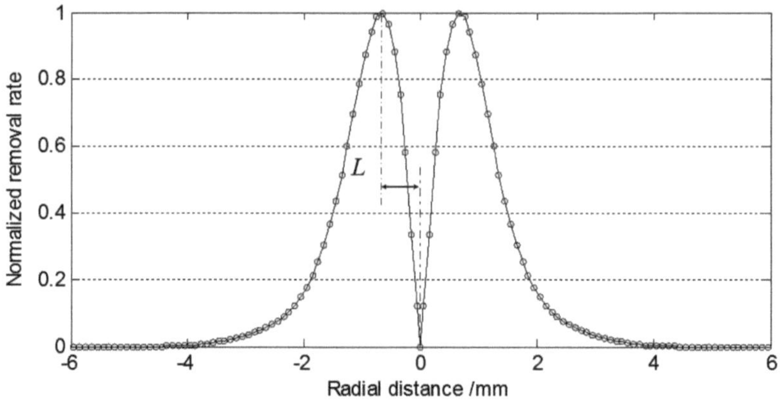

Fig. 2.26 Removal rate of normal jet polishing

Subsequently, the slurry is streaming radially away from the central point *O* along the surface, as seen in Fig. 2.25a. The normalized radial distribution of the rate of work done at the surface is calculated using commercially available Computational Fluid Dynamics (CFD) software. The model is shown in Fig. 2.25b. The diameter of the nozzle equals 1 mm, mesh area is 5 mm $\times$ 12 mm, and mesh size is 0.1 mm $\times$ 0.1 mm. The slurry jets form the inlet with a velocity of 20 m/s and flow out from the outlet (the boundary condition *P* equals 0, which is relative to the standard atmosphere), and the bottom of the mesh area is the wall. The velocity field distribution is shown in Fig. 2.25c.

By calculation through Eq. (2.26), the removal distribution with 1 mm diameter of the nozzle is exhibited in Fig. 2.26. The emulational result has a W-shaped profile across the polishing area. *L* is the horizontal distance between the peak and its adjacent valley, and it is about 0.7 mm when the diameter of the nozzle is 1 mm (see Fig. 2.26).

As seen in Fig. 2.26, the normal jet polishing does not have the deepest removal at the center. In order to obtain the Gaussian-like removal character, the eccentric rotation motion jet model is adopted. Figure 2.27 shows the polishing model with eccentric rotation motion. Figure 2.27a shows the principle of motion. The MR fluid with abrasives is delivered to the nozzle and directed vertically towards the workpiece. At the same time, nozzle spins on the axis of rotation with small eccentric distance *e* and angular-velocity. Figure 2.27b shows an arbitrary point *p* (r_2,θ_1) on the workpiece. *r* is the radius of the footprint around the central point *o*. The footprint of the normal jet along the trajectories forms a rotary symmetrical shape with the radius r_1 around the center of the rotation axis o_1. The distance between o_1 and *o* is eccentric distance *e*. The angle formed by $\overline{po}$ (length is *d*) and the *X*-axis is θ. When the jet rotates along the trajectories in a period, the effective range of polishing on *p* is $2\theta'$.

As shown in Fig. 2.28, rotation velocity vector $\mathbf{v}_e$ and fluid velocity vector $\mathbf{v}_l$ are introduced to describe relative velocity of the eccentric rotation motion mode on the

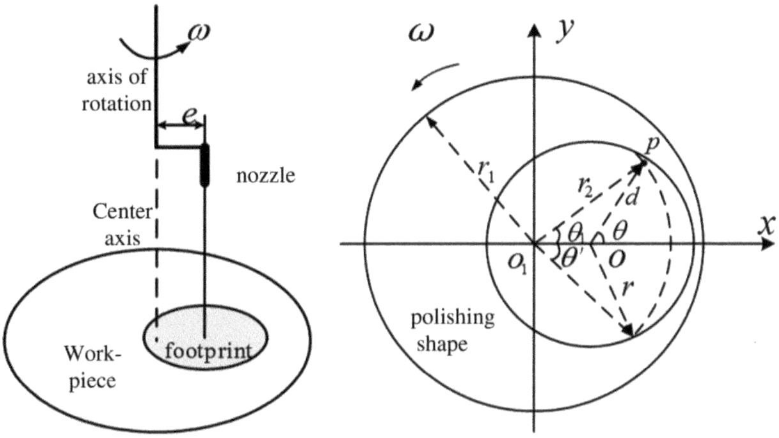

Fig. 2.27 Model of jet polishing with eccentric rotation motion. (**a**) Principle of motion; (**b**) polishing spot p on workpiece

Fig. 2.28 Relative velocity of the eccentric rotation motion model

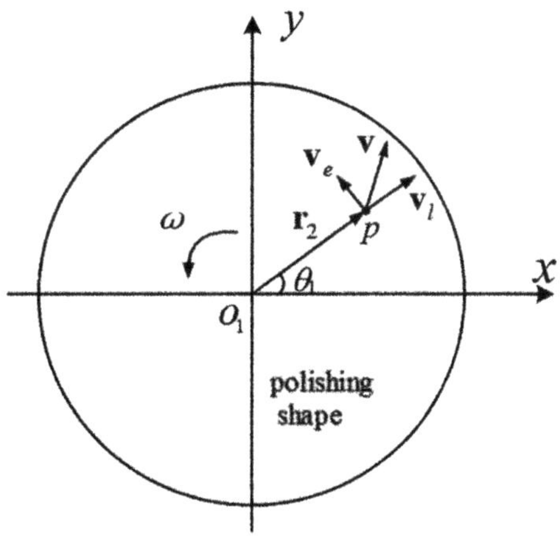

workpiece surface. The relative velocity at the point $p(\mathbf{r}_2,\theta_1)$ inside the polishing spot can be expressed as

$$\begin{cases} \mathbf{v} = \mathbf{v}_l + \mathbf{v}_e \\ \mathbf{v}_e = \boldsymbol{\omega} \times \mathbf{r}_2 \end{cases} \tag{2.27}$$

where $\mathbf{v}_e$ is caused by the nozzle's eccentric rotation motion, and $\mathbf{v}_l$ is the radial velocity due to impingement. When the diameter of the nozzle equals 1 mm, the eccentric distance e is 0.5 mm and $|\boldsymbol{\omega}|$ is 1 rps, computed $|\mathbf{v}_e|$ is 6.3×10^{-3} m/s, which is too small compared with the velocity of the jet ($|\mathbf{v}_l|$ is about 20 m/s), thus it can be ignored. Therefore, the $\mathbf{v}$ can be briefly described as $\mathbf{v}_l$.

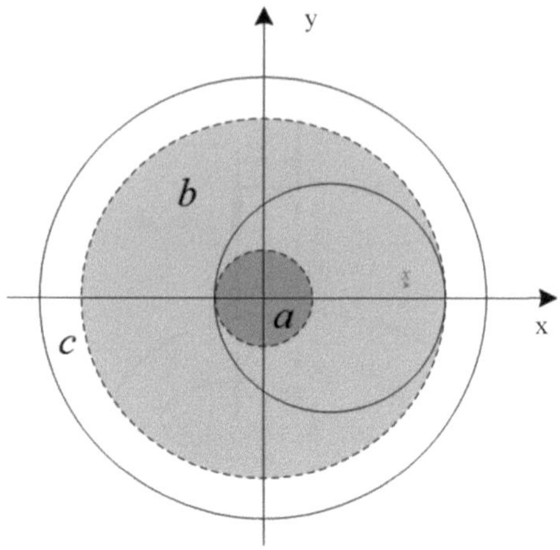

Fig. 2.29 Three areas in the polishing shape

Polishing shape can be expressed in three different areas, as shown in Fig. 2.29. The first area a is a circular shape with a radius $(r - e)$ around the point o_1, and footprint of the jet always covers on it in a period. The second area b is an annular shape, the radius of torus is from $(r - e)$ to $(r + e)$, and the effective range of polishing for an arbitrary point in this area is only the segment $(-\theta', \theta')$ in a period. The last area c is invalid. The material removal at the point $p(r_2,\theta_1)$ inside the polishing area in a period can be expressed as

$$R_1(r_2) = \int_{t_1}^{t_2} R(d,\theta)\mathrm{d}t = \frac{1}{\omega}\cdot\begin{cases} \int_{-\pi}^{\pi} R(d,\theta)\mathrm{d}\theta_1, & 0 \le r_2 \le r - e; \\ \int_{-\theta'}^{\theta'} R(d,\theta)\mathrm{d}\theta_1, & r - e < r_2 < r + e; \\ 0, & \text{others}. \end{cases} \tag{2.28}$$

Therefore, Eq. (2.28) can be described in a simple way:

$$R_1(r_2) = \frac{1}{\omega}\cdot\int_{-\bar{\theta}}^{\bar{\theta}} R(d,\theta)\mathrm{d}\theta_1 \tag{2.29}$$

where

$$\begin{cases} \bar{\theta} = \arccos\left\{\left[\left(r_2^2 + e^2 - r^2\right)/(2r_2 e)\cdot \mathrm{rect}[(r_2 - r)/(2e)] - \mathrm{step}\left(r - e - r_2\right)\right\}; \\ d = \left(e^2 + r_2^2 - 2er_2\cos\theta_1\right)^{1/2};\ 0 \le r_2 < r + e; \\ \theta = \arcsin(r_2\sin\theta_1/d). \end{cases}$$

2.3.3.2 Optimization of Parameters

Figure 2.30 shows the removal characters of magnetorheological jet polishing (MJP) with eccentric rotation motion. Figure 2.30a shows the normalized removal

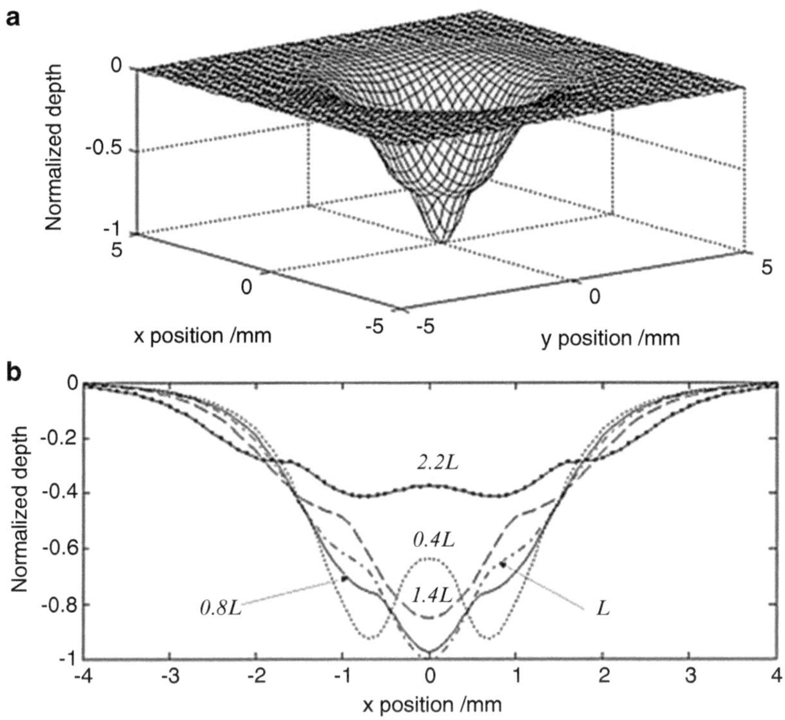

Fig. 2.30 Removal characters with eccentric rotation motion. (**a**) Removal shape with eccentric distance (L) and angular velocity (1 rps); (**b**) 2D profile sections across removal areas with different eccentric distances. *2D* two-dimensional

shape with eccentric distance L (as shown in Fig. 2.26) and angular velocity (1 rps), the removal shape has the maximum depth at the center and reduces to zero at the margin. Distribution of the removal shape possesses Gaussian-like character. The removal characters are then simulated with different eccentric distances ($0.4L$, $0.8L$, L, $1.4L$, and $2.2L$); results are shown in Fig. 2.30b. Sectional curve with eccentric distance ($0.4L$) shows W-shape, but removal depth is not zero and it rises dramatically in the center. Removal distributions with eccentric distances ($0.8L$, L, and $1.4L$) possess the character of Gaussian, but they have different depths and widths. Removal distribution is U-shaped when eccentric distance equals $2.2L$. Overall, the depth first increases then decreases and the width gradually increases, with the eccentric distance being larger.

In order to find out the excellent eccentric distance, a tending gene F is put forward to estimate the optimal eccentric distance. A higher value of F means the character closely approximates to the ideal Gaussian distribution. As described by Wang et al. [13], the F is expressed as below.

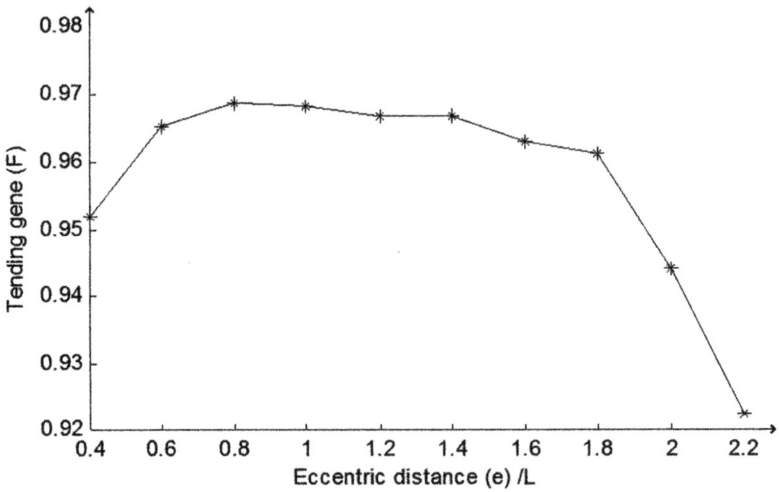

Fig. 2.31 The tending gene of material removal with different eccentric distance

$$F = \frac{W_{1/2}}{W} \tag{2.30}$$

where F is the tending gene, $W_{1/2}$ is the material removal in circular area with the half radius of the polishing spot around the center point, and W is the whole of the material removal.

For each different eccentric distance, the corresponding value of F is calculated and plotted in Fig. 2.31b. It is obvious that the distribution of the values is non-linear and unimodal distribution between $0.4L$ and $2.2L$, and the values of F are close to each other between $0.6L$ and $1.8L$. In particular, the maximum of F is 96.9 % (higher than 0.86 in the studies of Wang et al. [13]), when e is equal to $0.8L$. In addition, as seen in Fig. 2.30, the removal character possesses Gaussian-like character for $0.8L$. These indicate that $0.8L$ is the optimal eccentric distance and the removal character for $e = 0.8L$ is the optimal distribution.

2.3.3.3 Experimental Validation

Efforts have been made to find out the correlation between the experimental removal rate and the theoretical model with the self-made tool. K9 glasses with a diameter of 50 mm are used as the workpiece. The experimental settings are listed in Table 2.9.

With the above parameters, plenty of experimental studies with different distances have been conducted on the K9 glass with peak-to-valley (PV) 0.168 λ and root mean square (RMS) 0.031 λ ($\lambda = 632.8$ nm), as shown in Fig. 2.32.

Table 2.9 Parameters of polishing experiments

Parameter	Value
Mass fraction of CeO_2 particles in fluid (%)	6
Diameter of CeO_2 particle (μm)	2
Diameter of nozzle (mm)	1
Current (mA)	280
Number of turns in coil	4,000
Interior diameter of coil (mm)	6
Outside diameter of coil (mm)	18
High of coil (mm)	36
Diameter of wire (mm)	0.26
Pressure (MPa)	0.6

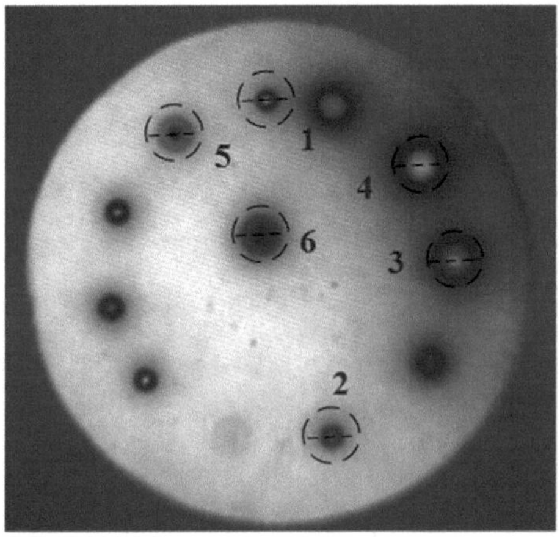

Fig. 2.32 Image showing the spots on the K9 glass

The six spots that are marked in Fig. 2.32 are the main objects of study. The first is the normal spot with the motionless tool. The remaining five spots show the shapes of polishing spots with eccentric rotation motion; their eccentric distances are about 0.4L, 0.8L, L, 1.4L, and 2.2L. The mask is necessary for researching the shape of the details, and the profiles across the removal areas are measured by ZYGO interferometer. Figure 2.33 shows the experimental results on the K9 glass for the profile curves of the polishing spots versus different eccentric distance.

Figure 2.33 shows the profile curves of the spots on K9 glass with a different jetting method (in each sub-graph, curve 1 and curve 2 indicate experimental and simulated data, respectively). Figure 2.33a shows the profile curve of the normal spot when the polishing tool maintains standstill. It has a peak in the middle and two valleys symmetrically at the adjacent margin. Besides, the curve accords with the distribution in Fig. 2.33, and its L equals approximately 0.7 mm. Figure 2.33b shows the degenerative W-shape in the second area when the eccentric distance is about 0.4L, the middle peak is reduced. The Gaussian-like curve would guarantee a successful polishing with the eccentric distance about 0.8L, which has maximum

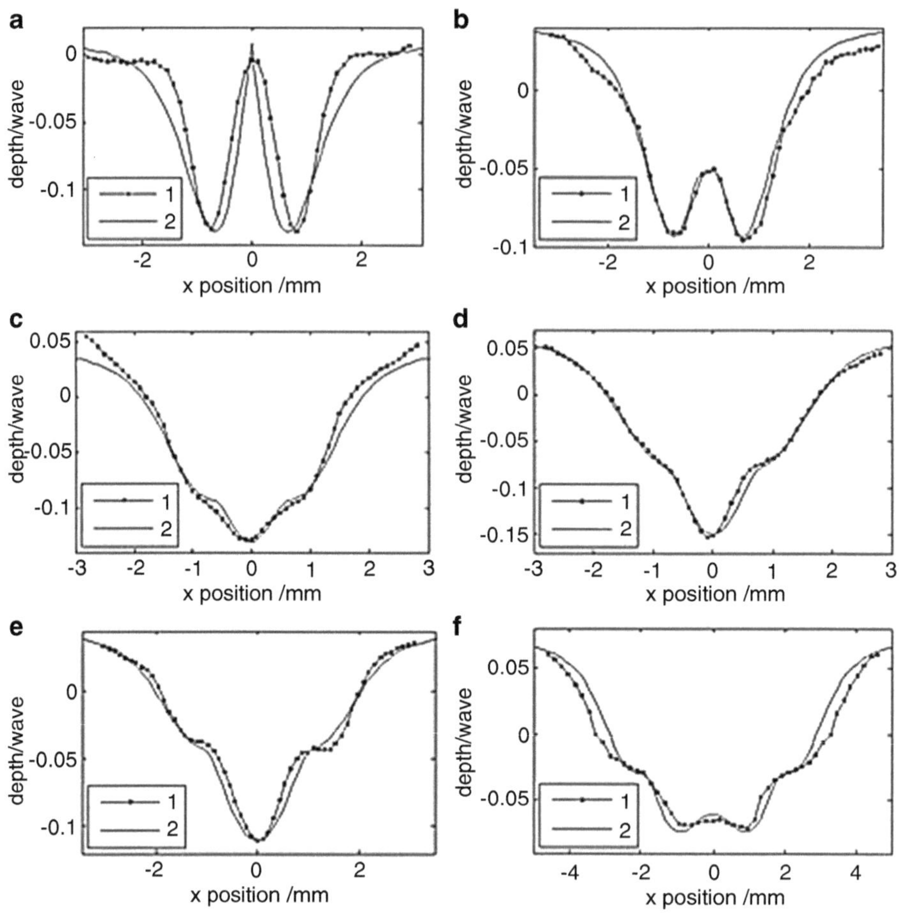

Fig. 2.33 The profile curves of the spots on K9 glass with different jetting method. (**a**) Normal spot; (**b**) eccentric distance is about 0.4L; (**c**) eccentric distance is about 0.8L; (**d**) eccentric distance is about L; (**e**) eccentric distance is about 1.4L; (**f**) eccentric distance is about 2.2L

removal at the center and reduces gradually, with the radius increasing. In particular, it is easy to observe that the curve varies smoothly, as shown in Fig. 2.33c. In Fig. 2.33d, the distribution of the profile is shown when the eccentric distance is about L, it is similar to the result of 0.8L, but there are step-shaped distributions between the center and margin. Step-shaped distributions are more obvious in Fig. 2.33e when the eccentric distance is about 1.4L. With the eccentric distance increasing, great changes have taken place in the distribution of the removal curve, as shown in Fig. 2.33f. It possesses not only the obvious step-shaped distributions but also a small protuberance in the middle.

According to above analysis, it is important to notice that in this case the profiles have been obtained in two different ways (experiment and simulation) resulting in the similar shapes.

2.3.4 *Electrorheological Finishing*

2.3.4.1 Model Constructions

Notwithstanding many years of investigation, there are still drawbacks in existing electrorheological (ER) polishing tools which restrict their practical applications. Traditional point-type tools contain only one of the electrodes in ER polishing. For a non-conductor workpiece, an auxiliary electrode needs to be placed close to the surface of the workpiece. This implies custom-made electrodes may be needed in some polishing work. Also, its location changes relative to the auxiliary electrode, when the tool is moved about onto the workpiece to polish. As the electric field changes with the tool location, this can result in a rather unstable removal and polishing performance.

ER fluids are usually non-conducting and have low viscosity. Viscosity increases significantly when an electric field is applied. ER fluids contain ER particles. These particles are perceived to form aggregates which align with the electric field, causing the ER fluid to change instantly from liquid to visco-plastic solid.

When an electric field is setup near the tip of a polishing tool and the tool is immersed in an ER fluid, the fluid within the influence of the field becomes viscous and effectively forms a small flexible polishing pad at the tool tip. With the addition of abrasive particles to the ER fluid, the abrasive particles are attracted to the ER particles in this virtual pad due to electrostatics. Under movement of the tool relative to the work surface, the abrasive particles are dragged through the gap between the tool and the surface, resulting in removal of asperities of the surface.

Within a certain range, field strength plays a dominant role in the viscosity of the ER fluid and in the attraction of abrasive particles to the ER particles. Thus, producing of a strong electric field at the tool tip is critical to the realization of ER polishing. An important application of ER polishing is in the polishing of small optics to high surface finishing. Quite often, such surfaces are non-conducting. They cannot be conveniently used as one of the electrodes in ER polishing [14]. An ER polishing tool with probe-like integrated-electrodes is proposed in this paper, aiming at the polishing of small non-conducting surfaces. The design concept is illustrated in the schematic diagram (Fig. 2.34). The central shaft (cathode) and the sleeve (anode) are exposed at the tool tip. ER fluid forms a virtual polishing pad near the tip when there is electrical potential difference between the two electrodes. Rotation of the central shaft causes the pad and the attached abrasive particles to also rotate and affects polishing action. There is no need for any additional electrodes for electric field formation.

A setup for ER polishing was built based on this design. The setup consists of a high-voltage power supply, a two-axis controller and a tool head (Fig. 2.35a). Key components of the tool head are the drive, translation device, and polishing device (Fig. 2.35b).

The device was firmly constructed. The three connecting rods and the upper and lower discs form the base frame of the polishing device. A motor on the upper disc is connected to the central shaft through a coupler.

A schematic of the lower part of the polishing device is given in Fig. 2.36a. A cone-shaped conducting sleeve is installed in the center of the lower disc, and the

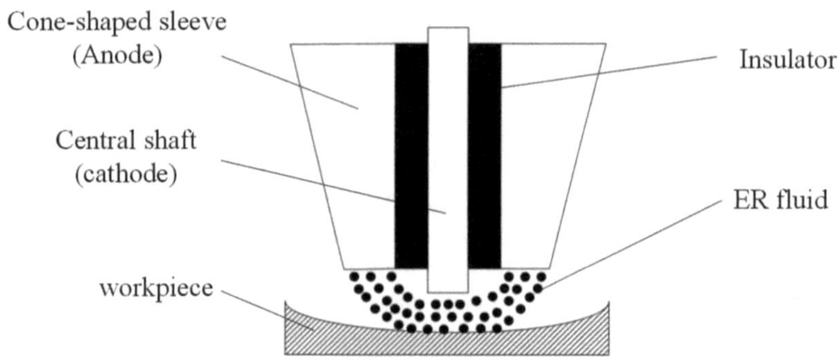

Fig. 2.34 Schematic diagram of design concept

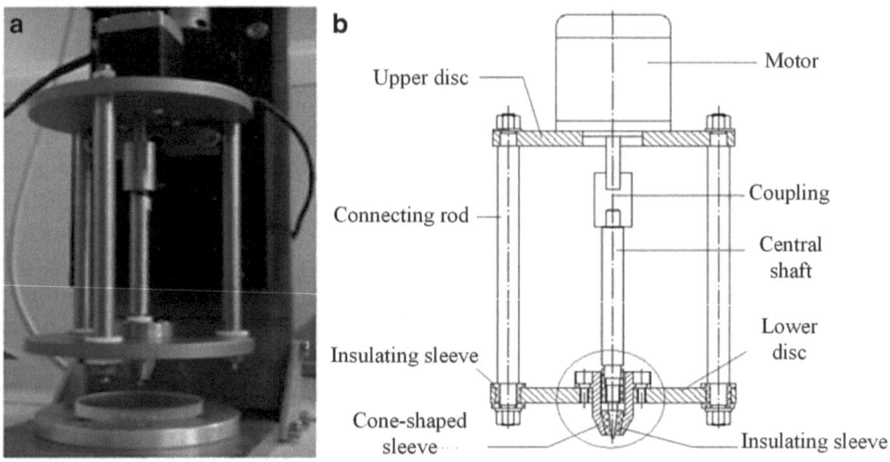

Fig. 2.35 ER polishing equipment. (**a**) Polishing device; (**b**) tool head

central shaft passes through the center of the cone-shaped sleeve. The central shaft serves as the cathode and is connected to the ground. The cone-shaped sleeve is the anode. An insulating sleeve is mounted on to the tip of the central shaft to prevent short circuiting when powered on. Rotary parts are accurately fabricated, well aligned, and supported to reduce vibration during processing.

In Fig. 2.36a, l is the distance of the central shaft beyond the cone-shaped sleeve, a is the diameter of the central shaft at the tip, and b and c are the inner and outer diameters of the cone-shaped sleeve. In the current setup, $l = 1$ mm, $a = 1.4$ mm, $b = 5.4$ mm, and $c = 8$ mm, respectively.

A photo of the tool head in action is shown in Fig. 2.36b. The tool head was positioned close to a specimen, and ER fluid was applied to the vicinity. Once the potential difference was established between the two electrodes, ER fluid became viscous and was seen attracted to the exposed part of the central shaft. An elevated ring of ER fluid was formed after the central shaft was set to rotate, which may be

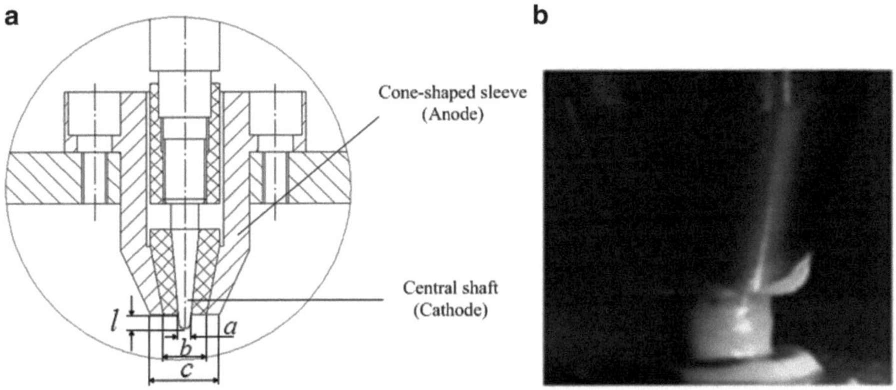

Fig. 2.36 Magnified views of the lower part of the polishing device. (**a**) schematic diagram; (**b**) polishing in process

Fig. 2.37 Coordinate system of the anode ring

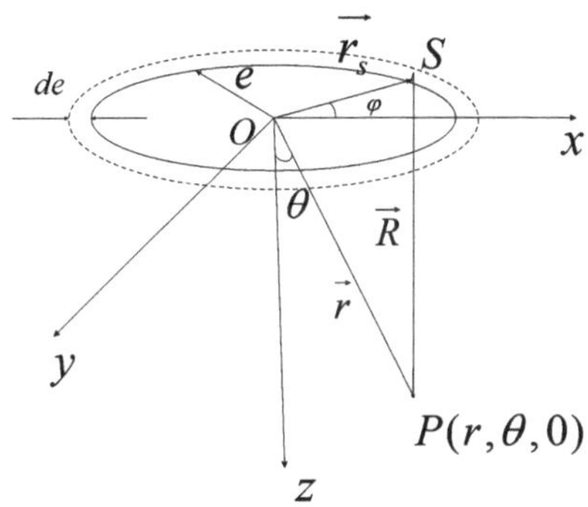

seen as a balance of gravitational force, centrifugal force, surface tension, and field-induced dipole attraction. The diameter of that ring was approximately 4 mm, slightly smaller than the inner diameter of the sleeve anode.

ER polishing effect is directly related to the strength of the electric field. In this section, the field strength near the tool tip is analyzed. The anode and the cathode are electrically charged when they are respectively connected to the two terminals of a voltage supply. The electric field strength in the vicinity of the tool tip depends on the shape and size of the anode and the cathode, and on the charge distribution. The field strength is analyzed based on Coulomb's law and making use of the superposition principle of electric fields. The effects of the anode and the cathode are considered separately and are then combined.

The bottom-end face of the anode is flat and annular (Fig. 2.37). Electric charge is assumed to be evenly distributed on the end face.

For the coordinate frame shown in the figure, the origin O is at the center of the end face and the Z-axis is along the centerline of central shaft. Consider a point $P(r,\ \theta,\ \Phi = 0)$ on the $x-z$ plane, with r, θ, and Φ being the radial distance, inclination angle, and azimuth angle, respectively. The position of P may be expressed as $\vec{r} = r\sin\theta\,\vec{i} + r\cos\theta\,\vec{k}$.

The end face of the anode may be seen as consisted of concentric rings. Consider the ring of radius e with infinitesimal width de. Let S be a small region of the ring at azimuth φ and the angle subtended at the origin is $\mathrm{d}\varphi$. The position of S is $\vec{r}_s = e\cos\varphi\vec{i} - e\sin\varphi\vec{j}$. The position of P relative to S is then given by:

$$\vec{R} = \vec{r} - \vec{r}_s = (r\sin\theta - e\cos\varphi)\cdot\vec{i} + e\sin\varphi\cdot\vec{j} + r\cos\theta\cdot\vec{k} \tag{2.31}$$

$$\left|\vec{R}\right| = \sqrt{e^2 + r^2 - 2er\sin\theta\cos\varphi} \tag{2.32}$$

If the charge density on the end surface of the anode is σ, the charge on S can then be expressed as $\mathrm{d}q = \sigma\cdot e\mathrm{d}\varphi\cdot \mathrm{d}e$. According to Coulomb's law, the electrical field strength at point P due to the charge on S is:

$$\begin{aligned}\mathrm{d}\vec{E} &= \frac{\vec{R}\,\mathrm{d}q}{4\pi\varepsilon\left|\vec{R}\right|^3}\\ &= \frac{\left[(r\sin\theta - e\cos\varphi)\cdot\vec{i} + e\sin\varphi\cdot\vec{j} + r\cos\theta\cdot\vec{k}\right]\sigma\cdot e\cdot \mathrm{d}e\cdot \mathrm{d}\varphi}{4\pi\varepsilon(e^2 + r^2 - 2er\sin\theta\cos\varphi)^{3/2}}\end{aligned} \tag{2.33}$$

where ε is the permittivity of the medium.

The electrical field at P due to the anode is:

$$\vec{E}_+ = \int_{b/2}^{c/2}\int_0^{2\pi} \mathrm{d}E = E_{x+}\vec{i} + E_{y+}\vec{j} + E_{z+}\vec{k} \tag{2.34}$$

where the three components of the field are, respectively:

$$E_{x+} = \frac{\sigma}{4\pi\varepsilon}\int_{b/2}^{c/2}\int_0^{2\pi} \frac{(x - e\cos\varphi)\cdot e\cdot \mathrm{d}e\cdot \mathrm{d}\varphi}{(e^2 + x^2 + z^2 - 2ex\cos\varphi)^{3/2}} \tag{2.35}$$

$$E_{y+} = \frac{\sigma}{4\pi\varepsilon}\int_{b/2}^{c/2}\int_0^{2\pi} \frac{e\sin\varphi\cdot e\cdot \mathrm{d}e\cdot \mathrm{d}\varphi}{(e^2 + x^2 + z^2 - 2ex\cos\varphi)^{3/2}} = 0 \tag{2.36}$$

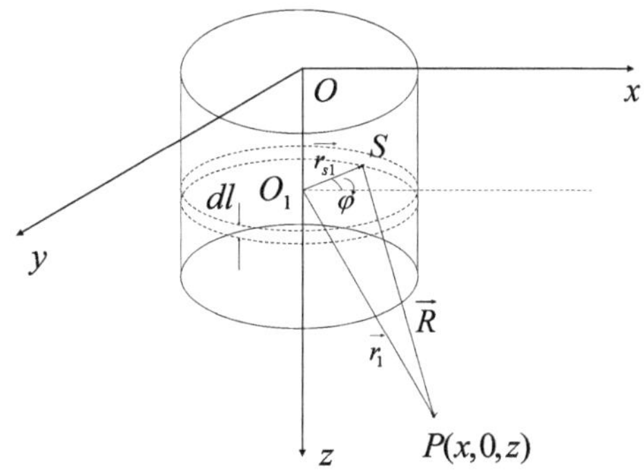

Fig. 2.38 Coordinate system of the cathode ring

$$E_{z+} = \frac{\sigma}{4\pi\varepsilon} \int\limits_{b/2}^{c/2} \int\limits_{0}^{2\pi} \frac{z \cdot e \cdot \mathrm{d}e \cdot \mathrm{d}\varphi}{\left(e^2 + x^2 + z^2 - 2ex\cos\varphi\right)^{3/2}} \tag{2.37}$$

The side of the central shaft which extends beyond the cone-shaped sleeve is approximately cylindrical (Fig. 2.38). Electric charge of the cathode is assumed to be evenly distributed on the surface of this extended part of the central shaft.

For the coordinate frame shown in the figure, the origin O is at the same level as the end face of the sleeve, and the Z-axis is along the centerline of the central shaft. Consider a point $P(x,0,z)$ on the x–z plane and a ring centered at O_1 with infinitesimal height dl. The position of P relative to the ring center is given by $r_1^{\rightharpoonup} = (x, 0, z - l)$. Let S be a small region of the ring at azimuth φ and the angle subtended at the O_1 is dφ. The position of S relative to O_1 is $r_{s1}^{\rightharpoonup} = \frac{a}{2} \cdot \cos\varphi \vec{i} - \frac{a}{2} \cdot \sin\varphi \vec{j}$. The position of P relative to S is then given by:

$$\vec{R} = \vec{r_1} - \vec{r_{s1}} = \left(x - \frac{a}{2}\cos\varphi\right)\vec{i} + \frac{a}{2}\sin\varphi\vec{j} + (z - l)\vec{k} \tag{2.38}$$

$$\left|\vec{R}\right| = \sqrt{x^2 + \left(\frac{a}{2}\right)^2 - ax\cos\varphi + (z - l)^2} \tag{2.39}$$

If the charge density on the cathode is η, the charge on S can then be expressed as $\mathrm{d}q = -\eta \cdot \frac{a}{2}\mathrm{d}\varphi \cdot \mathrm{d}l$. According to Coulomb's law, the electric field strength at point P due to the charge on S is:

$$\mathrm{d}\vec{E} = \frac{\vec{R}\,\mathrm{d}q}{4\pi\varepsilon\left|\vec{R}\right|^3} = \frac{-\left[\left(x - \frac{a}{2}\cos\varphi\right)\vec{i} + \frac{a}{2}\sin\varphi\vec{j} + (z - l)\vec{k}\right]\eta \cdot \frac{a}{2} \cdot \mathrm{d}\varphi \cdot \mathrm{d}l}{4\pi\varepsilon\left[x^2 + \left(\frac{a}{2}\right)^2 - ax\cos\varphi + (z - l)^2\right]^{3/2}} \tag{2.40}$$

The electrical field at P due to the cathode is:

$$\vec{E}_{-} = \int_0^l \int_0^{2\pi} \mathrm{d}\vec{E} = E_{x-}\vec{i} + E_{y-}\vec{j} + E_{z-}\vec{k} \tag{2.41}$$

where the three components of the field are, respectively:

$$E_{x-} = -\frac{a\eta}{8\pi\varepsilon}\int_0^l\int_0^{2\pi} \frac{\left(x - \frac{a}{2}\cos\varphi\right)\cdot \mathrm{d}\varphi \cdot \mathrm{d}l}{\left[x^2 + \left(\frac{a}{2}\right)^2 - ax\cos\varphi + (z-l)^2\right]^{3/2}} \tag{2.42}$$

$$E_{y-} = -\frac{a\eta}{8\pi\varepsilon}\int_0^l\int_0^{2\pi} \frac{\frac{a}{2}\sin\varphi\cdot \mathrm{d}\varphi \cdot \mathrm{d}l}{\left[x^2 + \left(\frac{a}{2}\right)^2 - ax\cos\varphi + (z-l)^2\right]^{3/2}} \tag{2.43}$$

$$E_{z-} = -\frac{a\eta}{8\pi\varepsilon}\int_0^l\int_0^{2\pi} \frac{(z-l)\cdot \mathrm{d}\varphi \cdot \mathrm{d}l}{\left[x^2 + \left(\frac{a}{2}\right)^2 - ax\cos\varphi + (z-l)^2\right]^{3/2}} \tag{2.44}$$

Because of the conservation of electric charge, the charge on the anode is equal to the charge on the cathode. Thus, one has:

$$\sigma \cdot s_{+} = \eta \cdot s_{-} \tag{2.45}$$

where $s_+ = \pi[(c/2)^2 - (b/2)^2]$ and $s_- = a \cdot l \cdot \pi$ are, respectively, the effective surface area of the anode and the cathode.

For the case of two parallel plates, the electric field is considered uniform and one has:

$$\frac{U}{d} = \frac{\sigma}{\varepsilon} \Rightarrow \sigma = \frac{U\varepsilon}{d} \tag{2.46}$$

where d is the distance between the plates and U is the potential difference. This is assumed to be applicable for the calculation of σ with $d = (b - a)/2$, which is the gap between the central shaft and the cone-shaped sleeve. The assumption of parallel plates is reasonably accurate below the sleeve and near the lower end of the central shaft where polishing takes place.

By the principle of superposition, the combined field:

$$\begin{aligned}\vec{E} &= E_x\vec{i} + E_y\vec{j} + E_z\vec{k}\\ &= (E_{x+} + E_{x-})\vec{i} + (E_{y+} + E_{y-})\vec{j} + (E_{z+} + E_{z-})\vec{k}\end{aligned} \tag{2.47}$$

$$E_x = \frac{U}{8\pi} \int_{2.7}^{4} \int_{0}^{2\pi} \frac{(x - e\cos\varphi) \cdot e \cdot d\varphi \cdot de}{(e^2 + x^2 + z^2 - 2ex\cos\varphi)^{3/2}} - \frac{0.544U}{\pi} \int_{0}^{1} \int_{0}^{2\pi} \frac{(x - 0.7\cos\varphi) \cdot d\varphi \cdot dl}{\left[x^2 + 0.49 - 1.4x\cos\varphi + (z - l)^2\right]^{3/2}} \tag{2.48}$$

$$E_y = 0 \tag{2.49}$$

$$E_z = \frac{U}{8\pi} \int_{2.7}^{4} \int_{0}^{2\pi} \frac{z \cdot e \cdot de \cdot d\varphi}{(e^2 + x^2 + z^2 - 2ex\cos\varphi)^{3/2}} - \frac{0.544U}{\pi} \int_{0}^{1} \int_{0}^{2\pi} \frac{(z - l) \cdot d\varphi \cdot dl}{\left[x^2 + 0.49 - 1.4x\cos\varphi + (z - l)^2\right]^{3/2}} \tag{2.50}$$

The field strength:

$$E = \left|\vec{E}\right| = \sqrt{E_x^2 + E_z^2} \tag{2.51}$$

When the point P is along the Z-axis, the electric field strength can be analytically solved as:

$$E = \frac{U \cdot z}{4} \left(\frac{1}{\sqrt{7.29 + z^2}} - \frac{1}{\sqrt{16 + z^2}} \right) - 1.089 \cdot U \left(\frac{1}{\sqrt{0.49 + (z - 1)^2}} - \frac{1}{\sqrt{0.49 + z^2}} \right) \tag{2.52}$$

Based on the above equation, changes of the field strength along the Z-axis with respect to the distance from the tip of the central shaft are shown in Fig. 2.39. The supply voltage varies from 3,000 to 1,500 V. The tip of the shaft is at $z = 1$ mm. The field strength is proportional to the supply voltage. It is strongest at the tip and decreases away from the tip. At a distance of 1 mm from the tip, the field strength is approximately half of it maximum value at the tip.

The electric field near the tool tip is also simulated using finite element analysis. Parameters used in the simulations are tabulated in Table 2.10. Simulation results are shown in Fig. 2.40. At the maximum supply voltage of 3,000 V, the maximum field strength is 2.46×10^6 V/m. The field strength decreases monotonically with the supply voltage. Field strength is particularly high near the edge of the central shaft and the inner edge of the end face of the sleeve, which can be explained by charge concentration at the edges. Yet, one can see that such effects diminish rather

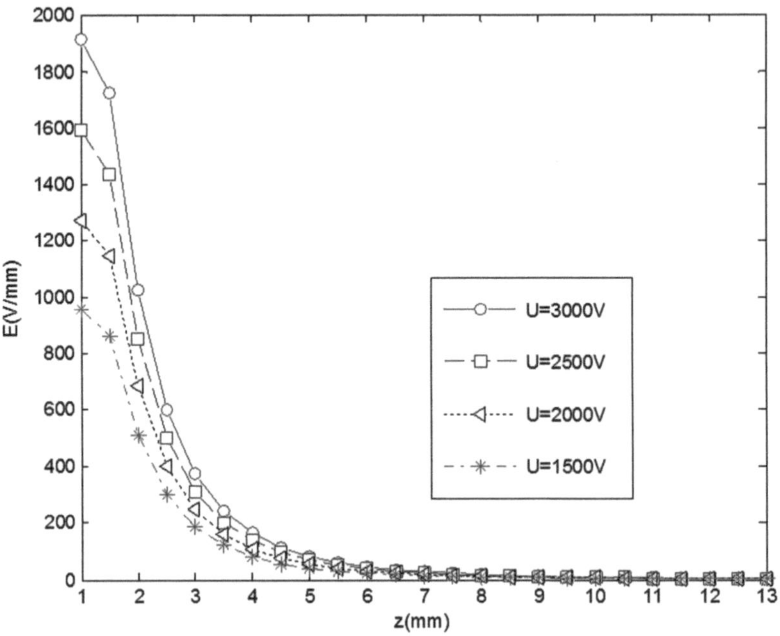

Fig. 2.39 Simulation of the electric field strength along the Z-axis

Table 2.10 Finite element analysis parameters

Material	Steel 0Cr18Ni9	Material resistivity	$9.7 \times 10^{-8} \Omega \cdot \text{m}$
Air resistivity	$10^6 \Omega \cdot \text{m}$	Cathode voltage	0 V
Anode voltage	1,500, 2,000, 2,500, 3,000 V		
Element type	Field analysis unit 67 with center of symmetry and far-field analysis unit 110		

rapidly from the edges and they are not noticeable beyond a small fraction of a millimeter from the edges. Overall, the field strength decreases quickly away from the tip of the central shaft, which is consistent with the analytical model of the electric field developed above and the results in Fig. 2.39. These also suggest the assumptions of uniform distribution of charge on the surface of the anode and the cathode are acceptable, and the analytical model provides reasonable prediction of the field strength as long as the point of interest is not within a small fraction of the edges.

2.3.4.2 Polishing Experiments

Experiments were conducted to examine the usefulness of the proposed tool with integrated electrodes for the polishing of non-conducting optics. The rate of

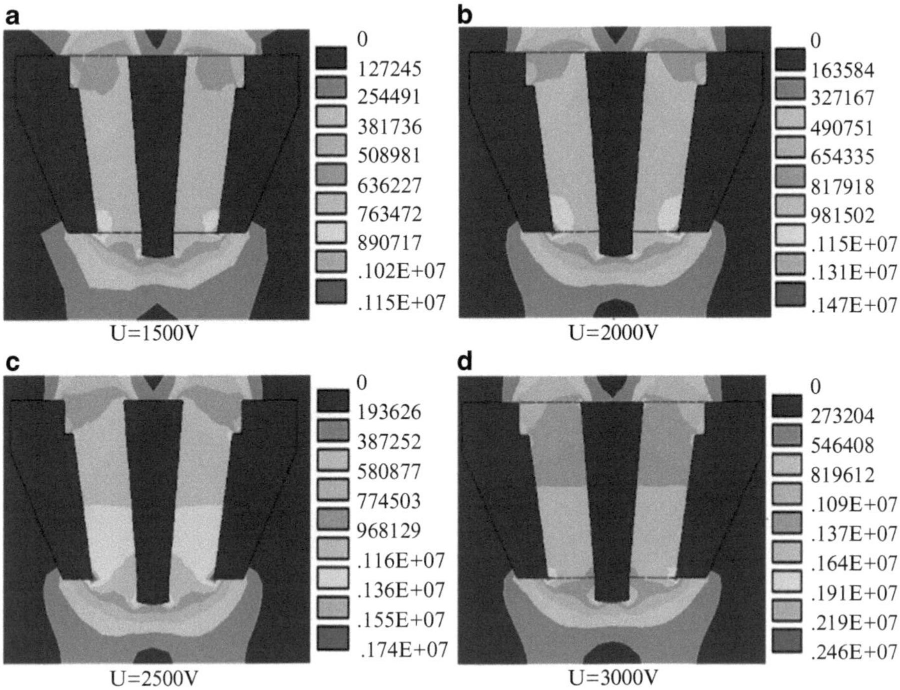

Fig. 2.40 Finite element analysis of electric field around the tool head

material removal is expected to increase with the field strength [11]. Two parameters to be studied in the current experiments are the supply voltage and the operating distance. In the last section, they were shown to have strong influence on the field strength. Their effects on ER polishing were investigated.

The ER fluid used in the experiments was composed of 47.62 % starch, 47.62 % silicone oil, and 4.76 % ceria. For all experiments, the rotating speed of the central shaft was set at 1,500 rpm and the polishing time was set at 30 min. The tool was perpendicular to the specimen. The specimens were made of K9 glass and were prepared using traditional grinding and polishing to a level of finishing suitable for the experiments.

The supply voltage was 2,000 V, and the operating distance varied between 0.5 and 0.8 mm. The surface morphology of the specimens was measured before and after ER polishing with a FISBA-μ820 interferometer. The measured area was set at 10.7 mm × 9.6 mm, which was slightly smaller than the impression on the specimen by the tool. The CCD resolution of 1,024 × 1,020 pixels was suitable for form or waviness measurement or surface irregularities investigation over a larger range.

The captured surface profiles are shown in Fig. 2.41. Polishing leads to material removal. There is clear evidence that such removal in ER polishing can smooth the surface profiles and reduce the PV values.

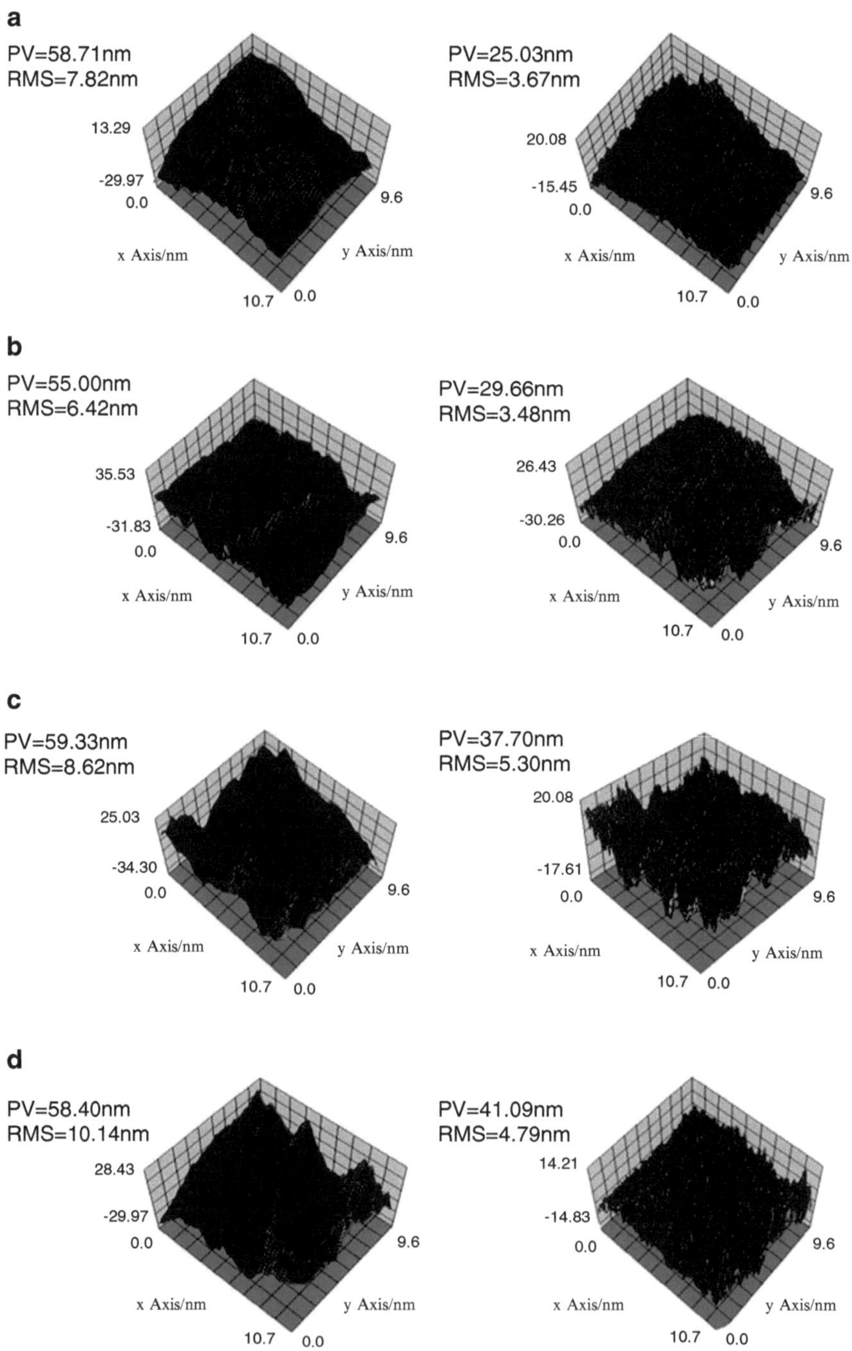

Fig. 2.41 Changes of the surface profile ($U = 2{,}000$ V). (**a**) $z = 0.5$ mm; (**b**) $z = 0.6$ mm; (**c**) $z = 0.7$ mm; (**d**) $z = 0.8$ mm

Table 2.11 Peak-to-valley versus z

z (mm)	Before (a_1) (nm)	After (a_2) (nm)	$a_1 - a_2$ (nm)	K (%)
0.5	58.71	25.03	33.68	57.37
0.6	55.00	29.66	25.34	46.07
0.7	59.33	37.70	21.63	36.46
0.8	58.40	41.09	17.31	29.64

One can observe more obvious polishing effect when the operating distance is shorter. The final PV is smaller and the reduction in PV is larger when the operating distance is shorter (Table 2.11). It is perceived that the ER fluid forms a virtual polishing pad around the tool tip, under the influence of the electric field. The stronger effects on the PV reduction when the operating distance is shorter suggest that the stiffness of the virtual pad increases closer to the tip of the tool. The increase in stiffness is attributed to the higher viscosity of the ER fluid closer to the tool tip because of the increase in field strength near the tip. The greatest reduction in PV obtained for the operating distance was 0.5 mm where the PV was reduced from 58.71 to 25.03 nm.

The relative change of an indicator is expressed as:

$$K = \frac{a_1 - a_2}{a_1} \tag{2.53}$$

where a_1 and a_2 are the values of the indicator before and after processing, respectively.

The relative change of PV is plotted against the operating distance in Fig. 2.42. The monotonic trend of the plot suggests the removal of surface irregularities is more effective and faster when the tool is closer to the specimen.

RMS of a surface profile gives more averaged measurement (Table 2.12). The relative change of RMS is also plotted against the operating distance in Fig. 2.42. The result also suggests a monotonic trend, except the case at the operating distance is 0.8 mm. Inspection of the surface profiles in Fig. 2.41 reveals that more material was indeed removed for the cases of the operating distance from 0.5 to 0.7 mm. The specimen for the 0.8 mm operating distance contained relatively more narrow ridges and grooves on the surface. It was likely that, for that specimen, a larger reduction in RMS could be achieved with less removal of material.

The operating distance was set at 0.5 mm, and the supply voltage varied between 1,500 and 3,000 V. The surface roughness Ra of the specimens was measured before and after ER polishing with a wyko NT1100 interferometer. The measured area covered 227 μm × 298.5 μm. The CCD resolution of 736 × 480 pixels was adequate for roughness measurement or surface irregularities investigation over a smaller range.

Comparison of the surface profile before and after polishing was shown in Fig. 2.43. Results indicate successful reduction of the surface roughness Ra in the nanometer range. In this range, resulting surface profiles did not resemble the original profiles. The Ra of the original surfaces was between 4.1 and 8.5 nm

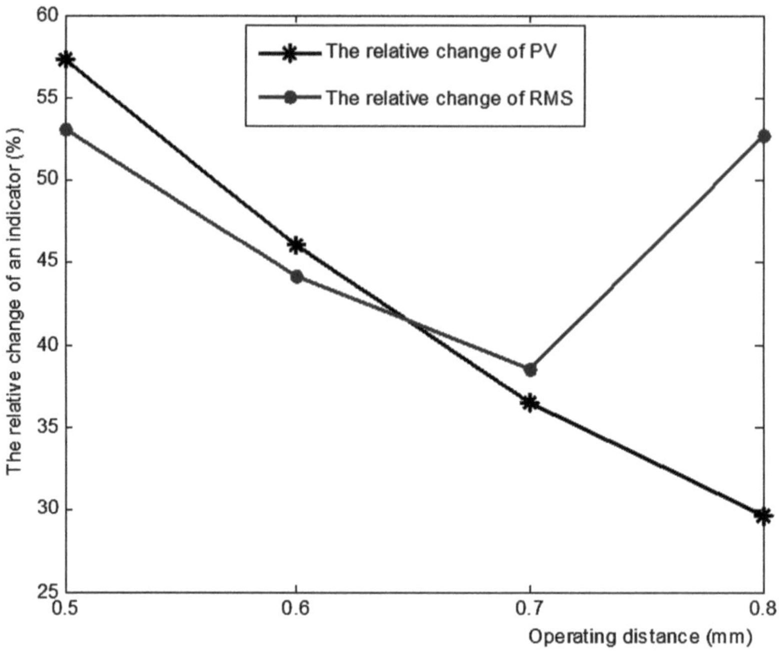

Fig. 2.42 Relationship between the relative change of PV and RMS and the operating distance (U = 2,000 V). *PV* peak-to-valley, *RMS* root square mean

Table 2.12 Root mean square versus z

z (mm)	Before (a_1) (nm)	After (a_2) (nm)	$a_1 - a_2$ (nm)	K (%)
0.5	7.82	3.67	4.15	53.07
0.6	6.24	3.48	2.76	44.23
0.7	8.62	5.30	3.32	38.52
0.8	10.14	4.79	5.35	52.76

(Table 2.13). The Ra of the resulting surfaces was between 2.5 and 2.8 nm, with the exception of the case of U = 2,000 V where Ra was 3.9 nm. These suggest that, within the voltage range, the supply voltage did not play a significant role in the achieved Ra. For the case of U = 2,000 V, one can see a long and broad ridge across the surface, which may be the reason for the higher Ra of 3.9 nm. A likely explanation is inhomogeneity of the specimen as all the other specimens seem to have rather unremarkable surface profiles after polishing.

The relative change of Ra is plotted against the supply voltage in Fig. 2.44. The effect on the rate of Ra reduction seems to be larger with a larger supply voltage. Relatively speaking, the effect is more distinct when the voltage is increased from 1,500 to 2,000 V, compared with the increase from 2,000 to 2,500 V. It is about to level off when the voltage is beyond 3,000 V.

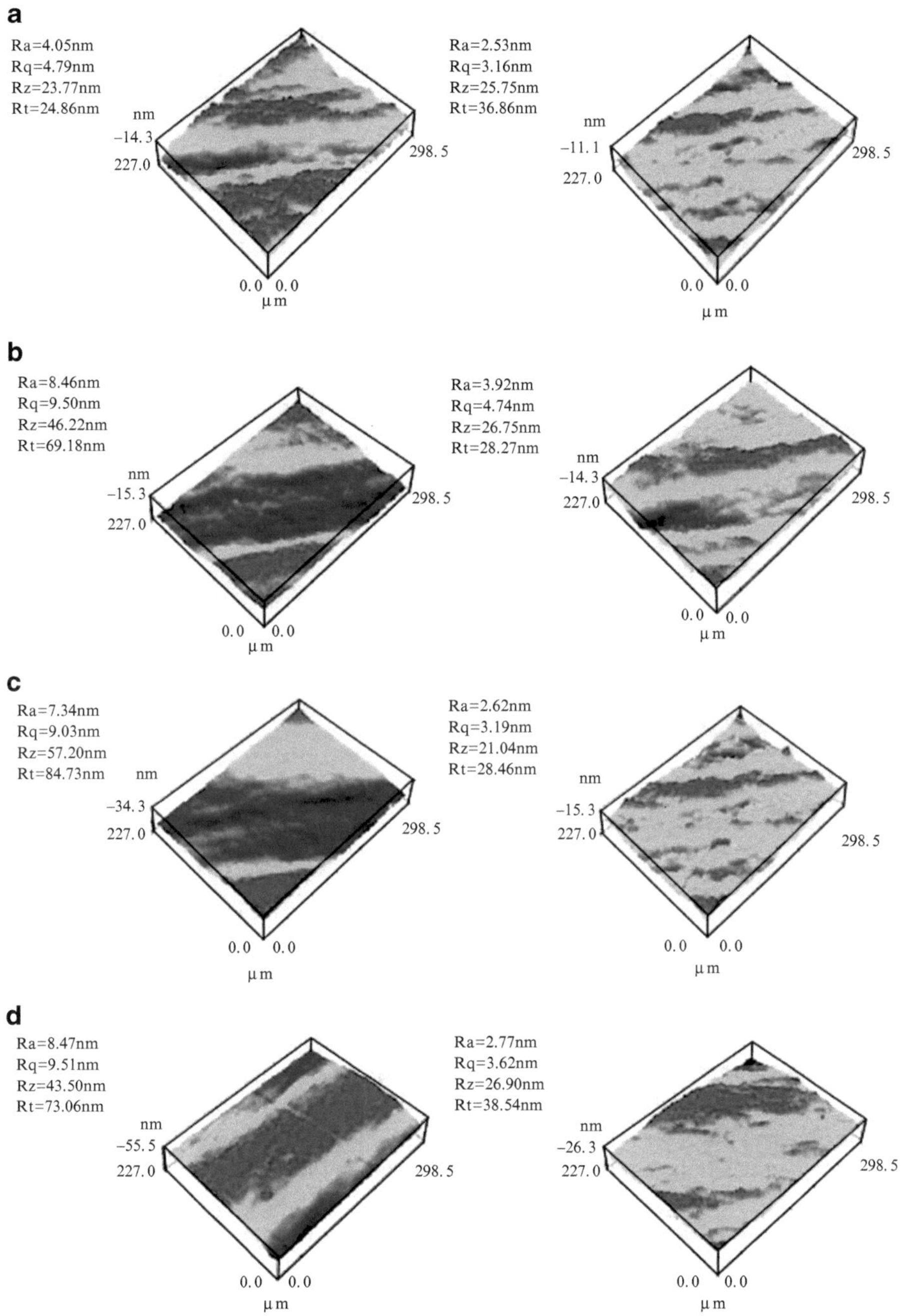

Fig. 2.43 Changes of the surface roughness (operating distance 1.5 mm)

Table 2.13 Roughness average (Ra) versus *U*

U (V)	Before (a_1) (nm)	After (a_2) (nm)	$a_1 - a_2$ (nm)	*K* (%)
1,500	4.05	2.53	1.52	37.53
2,000	8.46	3.92	4.54	53.66
2,500	7.34	2.62	4.72	64.31
3,000	8.47	2.77	5.70	67.30

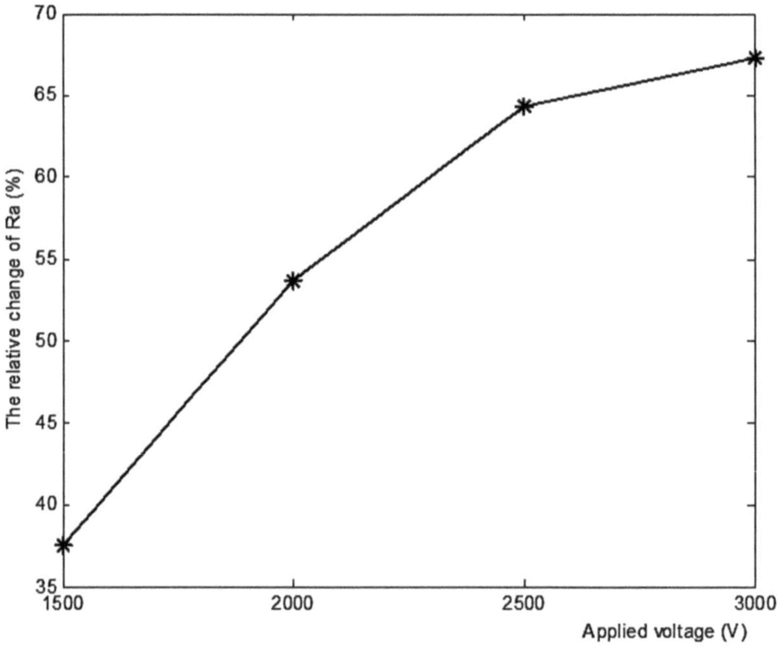

Fig. 2.44 Relationship between the relative change of roughness average (Ra) and the supply voltage (operating distance 0.5 mm)

Presence of abrasive particles in the ER fluid is important to material removal and Ra reduction. The rate of removal depends on the amount of abrasive particles in contact with the surface. A larger supply voltage seems to induce better attraction of the abrasive particles to the ER particle aggregates. Thus, higher rates of removal are obtained when the aggregates are rubbed against the surface, as they are carried by the rotation of the central shaft.

Leveling off of the curve in Fig. 2.44 indicates that further increase in the supply voltage cannot lead to further significant increase in removal rate. It is possible that further voltage increase did not help to attract additional abrasive particles. It is also possible that there was no further formation of ER particle aggregates.

In traditional polishing, the final Ra of a surface depends mainly on the size of the abrasive particles. To a lesser extent, it also depends on the force on the polishing tool. In the current experiments, the final Ra did not vary notably with the supply voltage. This is reasonable as the same type of abrasives was used in all the experiments. Although the supply voltage affected the viscosity of the virtual pad and thus, to some extent, the indentation of abrasive particles, the effects on the Ra were negligible.

References

1. Juranek HJ, Sand R, Schweizer J, Harnisch B, Kunkel BP, Schmidt E, Litzelmann A, Schillke F, Dempewolf G (1998) Off-axis telescopes: the future generation of Earth observation telescopes. In: SPIE international symposium on optical science, engineering, and instrumentation. International Society for Optics and Photonics, Bellingham, pp 104–115
2. Rupp V (1965) The development of optical surfaces during the grinding process. Appl Opt 4(6):743–748
3. Jones RA (1977) Optimization of computer controlled polishing. Appl Opt 16(1):218–224
4. Cheng H, Feng Z, Wu Y (2004) Process technology of aspherical mirrors manufacturing with magnetorheological finishing. Mater Sci Forum 471–472:6–10
5. Cheng H, Yeung Y, Wang Y (2009) Experimentation on MR fluid using a 2-axis wheel tool. J Mater Pro Tech 209(12-13):5254–5261
6. Golini D (1999) Precision optics manufacturing using magnetorheological finishing (MRF). In: Singapore. International Society for Optics and Photonics, Bellingham, pp 78–85
7. Cheng H, Feng Z, Wang Y, Lei S (2005) Magnetorheological finishing of SiC aspheric mirrors. Mater Manuf Process 20(6):917–931
8. Chi C, Wang Z, Zhao P (1993) Iron magneto-fluid-mechanics. BeiHang University Press, Beijing, pp 378
9. Cheng HB, Wang T, Feng YP, Yam Y (2011) Process planning and tool design of magnetorheological jet finishing. Appl Mech Mater 44:222–226
10. Tricard M, Kordonski W, Shorey A, Evans C (2006) Magnetorheological jet finishing of conformal, freeform and steep concave optics. CIRP Ann Manuf Technol 55(1):309–312
11. Kordonski WI, Shorey AB, Tricard M (2006) Magnetorheological jet (MR Jet™) finishing technology. J Fluids Eng 128(1):20–26
12. Becker H, Hottel H, Williams G (1967) The nozzle-fluid concentration field of the round, turbulent, free jet. J Fluid Mech 30(2):285–303
13. Wang T, Cheng H, Dong Z, Tam H (2013) Removal character of vertical jet polishing with eccentric rotation motion using magnetorheological fluid. J Mater Pro Tech 213(9):1532–1537
14. Cheng H-B, Feng Z-J, Cheng K, Wang Y-W (2005) Design of a six-axis high precision machine tool and its application in machining aspherical optical mirrors. Int J Mach Tool Manuf 45(9):1085–1094

Chapter 3
Finishing Paths

Abstract This chapter introduces various finishing paths and corresponding generating rules. In particular, we present the Peano-like path and investigate its superior advantages in deterministic figuring or uniform smoothing processes. Analysis and experiments were conducted to validate the influence of different paths and path pitch to the convergence rate and fabrication efficiency, as well as mid-frequency errors and surface roughness.

3.1 Overview of Finishing Paths

Tool paths are indispensable for uniform or deterministic optical surfacing systems, as well as lots of other computer numerical control (CNC) fabrication machines. A polishing tool can be driven by machines to traverse all over the optical surface with varying velocities for reducing surface roughness, controlling surface form, and diminishing subsurface damage. Paths are so significant to deterministic polishing that they need to be further investigated for increasing the surface quality, especially the surface texture and mid-spatial frequency errors.

A basic requirement of tool paths for optical surfacing systems is that the surface can be completely covered during a polishing cycle. Based on this, a scanning path has been widely adopted in many surfacing applications, which involves a raster, spiral, and concentric circle, etc. The tool can follow a scanning path continuously in its feeding direction but has a saltation (i.e., scanning interval) in the vertical direction (namely, the scanning direction). Thus, the scanning interval should be selected carefully according to the shape and size of the tool influence functions (TIFs). If the scanning interval was set too large, it would inevitably result in grooves and ripples (or waviness) on the surface; these are also known as mid-spatial frequency errors or ripples.

With the exception of the scanning path, Mizugaki et al. [1] claimed that Peano paths could be advantageous in polishing because tool path directions are more balanced on the surface compared with scanning paths. Bi-scanning [2] and

H. Cheng, *Independent Variables for Optical Surfacing Systems: Synthesis, Characterization and Application*, DOI 10.1007/978-3-642-45355-7_3,

extended scanning [3] were proposed for achieving more even coverage of free-form surfaces in polishing. There were also attempts to mimic the polishing action of the human hand. Cho et al. [4] experimented with Lissajous curves to reduce waviness introduced onto the surface by polishing. Pessoles and Tournier [5] used trochoidal curves to cover the surface in a multidirectional manner. Dunn and Walker [6] also presented a pseudo-random tool path to cover a hexagon region for improving surface texture and roughness. Hu et al. [7] adopted a random pitch to suppress ripple errors. Deng et al. [8] used a weighted iterative algorithm to design a path according to the surface error distribution. These paths have partly improved the polishing performance and have some advantages compared with scanning paths.

In this chapter, the path design detailed is investigated and a series of path generation rules are presented, including scanning, fractal, boundary-adaptive, and surface form-based paths. In particular, a Peano-like path is investigated in detail by simulations and experiments, which have validated the features of multiple directions and has better smoothing effects for mid-spatial frequency errors and surface texture.

The influence of the path interval on convergence rate and mid-spatial frequency errors are simulated and feasibility validations are carried out, respectively; these conclude that the path interval has better performance as it is reduced; in the mean time, fabrication efficiency would not be influenced.

3.2 Multiple Tool Path Mode

3.2.1 Raster Paths Mode

Raster is the most common path used in CNC fabrication machines. The surfacing tool feeds linearly from one side of the workpiece to the other end and then feeds to a vertical direction for a pitch, known as the path interval. This process is repeated so that the whole surface is covered. Figure 3.1 presents raster paths covering a square and hexagon region. The feeding direction is vertical to the scanning direction. CNC machines can achieve this path mode as long as it has two orthogonal linear axes.

The path is easy to generate and can cover any region with a closed boundary. Moreover, it can be used in localized high regions of surfaces. As shown in Fig. 3.2, the surface form has a high region assigned by a circle. If we use the jet polishing tool to polish this region, we can set a raster path, as shown in Fig. 3.3b. The polishing points are all located in the high region, and are amplified, as shown in Fig. 3.3a. The result after jet polishing is presented in Fig. 3.2b, which indicates that the high region of the surface form was diminished a lot.

It is easy to understand that the tool can make a continuous motion along the feeding direction with constant or varying velocities according to the setup of the dwell time. However, the motion in the scanning direction is discontinuous, which is

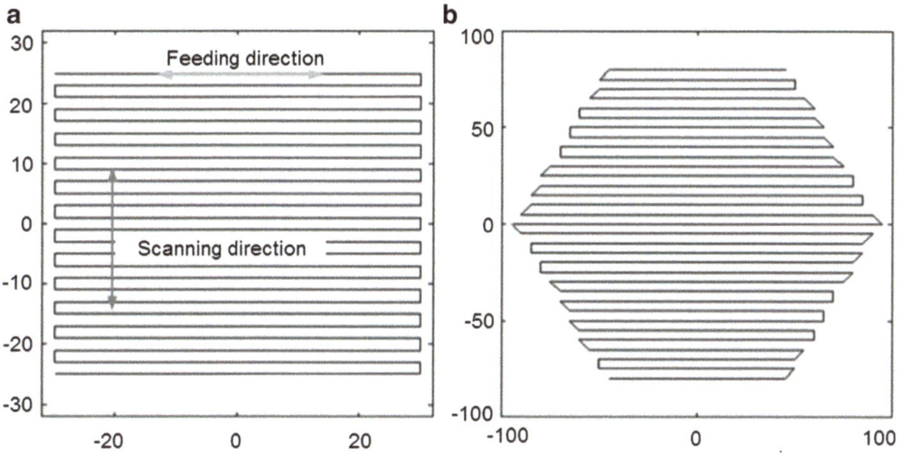

Fig. 3.1 Raster path covers a (**a**) 25 mm × 25 mm square and (**b**) hexagon with 200 mm across size

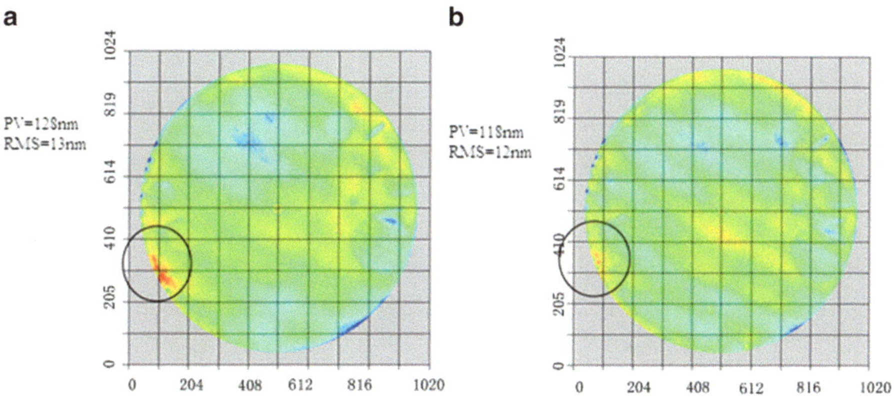

Fig. 3.2 Surface forms (**a**) before jet polishing and (**b**) after local jet polishing

discretized by the path interval. The superpositions of TIF footprints would generate grooves or waviness, which also belong to mid-spatial frequency errors.

3.2.2 *Spiral Paths Mode*

3.2.2.1 How to Generate a Spiral Path

The polar coordinate fabrication regime is widely used in optical surfacing systems. It drives the tool traversing a radius from the inner to the outer direction while a turntable is continuously rotating 360°. For a polishing process using polar

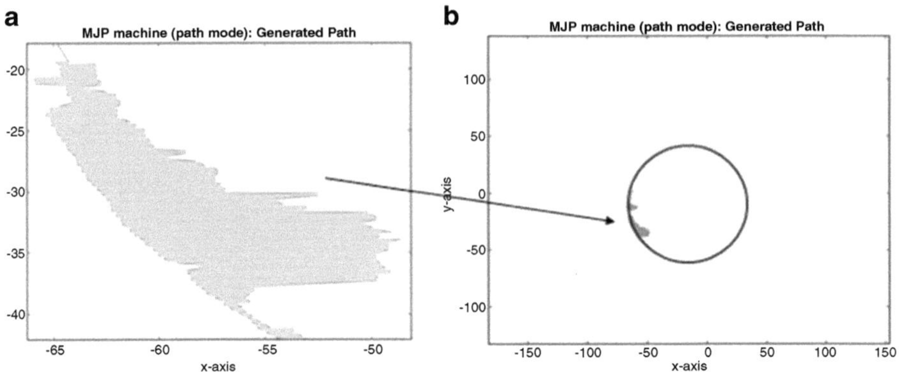

Fig. 3.3 Localized raster tool path covers the high region of the surface form. (**a**) amplified area; (**b**) overall view

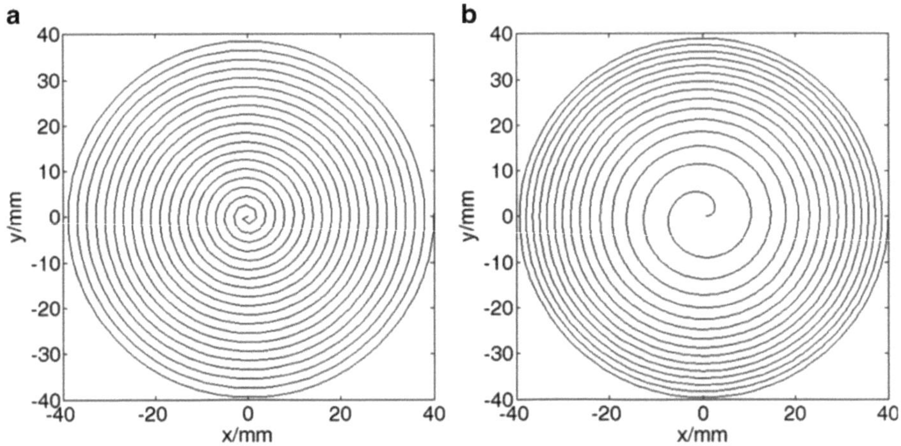

Fig. 3.4 Spiral polishing path. (**a**) Uniform pitch, $\beta = 1$; (**b**) Non-uniform pitch, $\beta = 0.5$

coordinates, the common Archimedes spiral path (shown in Fig. 3.4a) with identical pitch is usually adopted. It can be denoted as Eq. (3.1):

$$\rho = b \cdot \theta \tag{3.1}$$

where (ρ, θ) represents the position of the polishing path in the polar coordinate, and b indicates a constant proportionality coefficient that correlates with the screw pitch (SP):

$$\mathrm{SP} = 2\pi b \tag{3.2}$$

where b is a constant that indicates that the SP is equal all over the path. In the polar coordinate fabrication process, to reduce the desired velocity in the central region, the dwell time in the central points should be increased by

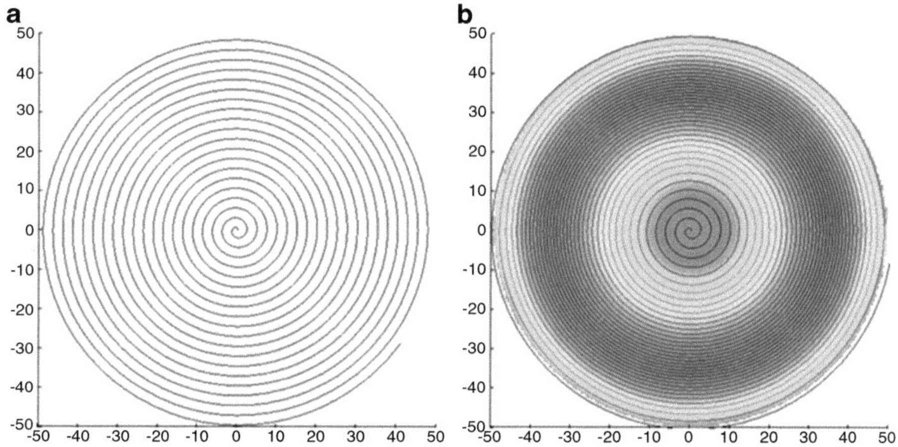

Fig. 3.5 Spiral paths of (**a**) uniform screw pitch and (**b**) non-uniform path

reducing the number of central points. As a method, the SPs of a spiral path could be diminished by increasing radial distance; ρ can be defined as in Eq. (3.3).

$$\rho = b \cdot \theta^{\beta} \tag{3.3}$$

where β represents a path pitch factor. If $\beta > 1$, the SP would increase as ρ increases; $\beta < 1$ denotes that the SP would decrease as ρ increases; $\beta = 1$ is the Archimedes spiral. Set as $\beta = 0.5$, the path is plotted in Fig. 3.4b. The path density of the central region is much lower than that of the edge region.

In addition, b can be set to be a position-variant factor related to the distribution of surface form. The path can gather more polishing points on high regions of the surface form, which is useful when removing rotational symmetric errors on the surface. Define a surface form as $Z_{m\times n}$, which is related to a Φ100-mm workpiece, and an initial pitch factor of b_0, and normalize $Z_{m\times n}$ into a range of 0–1. Regarding the ith path point, its position can then be confirmed by Eq. (3.4):

$$\begin{cases} b_i = b_0(1 - Z_{i-1}) \\ \rho_i = b_i\theta \end{cases} \tag{3.4}$$

As an example, shown in Fig. 3.5, a uniform and nonuniform path covers a Φ100-mm circle. The SP of the right path is position-variant according to the error distribution of the surface form. Following this path, the tool would dwell in the higher girdle band with more time to remove more material.

3.2.2.2 How to Get Dwell Points from Spiral Paths

There are two frequently used methods that can get dwell points from the spiral path: equal angle interval and equal arc length. With respect to the former method,

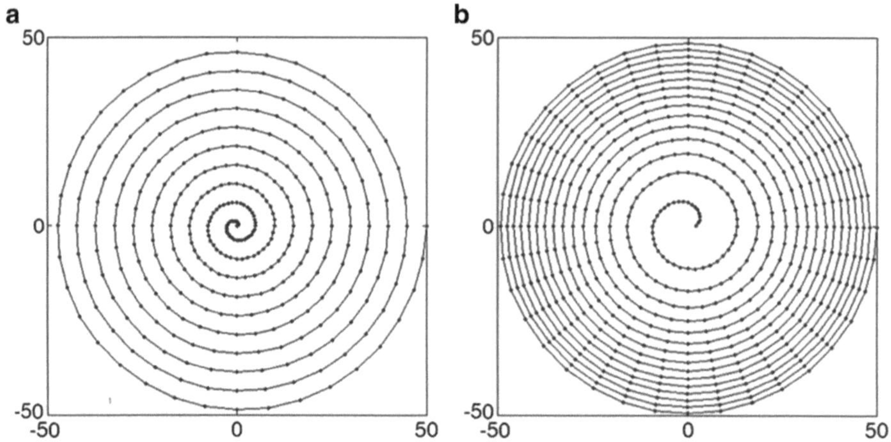

Fig. 3.6 Equal angle spiral paths with (**a**) uniform pitch and (**b**) non-uniform pitch

the intersection angles of any two adjacent dwell points are equal all over the surface. The polar coordinate of the *i*th dwell point can be expressed as:

$$\begin{cases} \theta_i = id\theta \\ \rho_i = b\theta_i \end{cases}, \tag{3.5}$$

where $d\theta$ denotes the interval of the intersection angle. The method is suitable for any kind of spiral path, whatever the uniform or nonuniform spiral. The example shown in Fig. 3.6a indicates a uniform spiral path with an equal angle interval in which $b = 0.796$, $d\theta = 10\ ^\circ$. Figure 3.6b is the nonuniform spiral path with an equal angle interval: $\beta = 0.5$, $d\theta = 10\ ^\circ$, and $b = 5.15$.

It is easy to find that the arc length of two adjacent dwell points increases monotonously as θ increases. The uniformity for covering surfaces is lower, which may result in nonuniform material removal and mid-spatial frequency errors. As a result, this kind of path is rarely used in deterministic polishing of optical mirrors.

The second method is called the equal arc length method (Fig. 3.7); it means the arc length between two adjacent points is generally equal. We perform analysis for uniform pitch spiral and non-uniform pitch spiral. With respect to uniform pitch spiral we can get the data set of θ from Eqs. (3.6) and (3.7). Where arc _ length represents the pre-defined arc length.

$$2\text{arc\_length} = (\rho_{i+1} + \rho_i)(\theta_{i+1} - \theta_i) \tag{3.6}$$

$$\theta_{i+1} = \sqrt{{\theta_i}^2 + 2\text{arc\_length}/b} \tag{3.7}$$

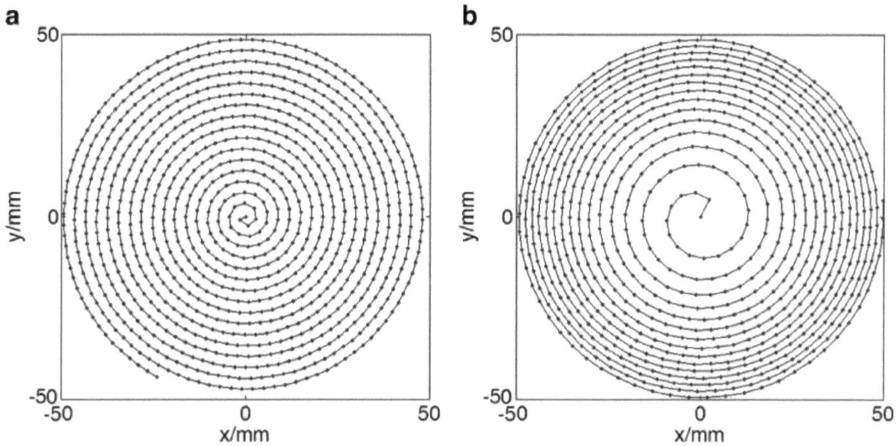

Fig. 3.7 Equal arc spiral path of (**a**) uniform pitch spiral and (**b**) non-uniform spiral

However, the situation for the non-uniform spiral is more complex, which is mainly caused by the value of β. We regard $\beta = 0.5$, and combine Eqs. (3.3) and (3.6):

$$\theta_{i+1}{}^{\frac{3}{2}} - \theta_i \theta_{i+1}{}^{\frac{1}{2}} + \theta_{i+1}\theta_i^{\frac{1}{2}} - \theta_i^{\frac{3}{2}} = 2\text{arc\_length}/b \tag{3.8}$$

Assuming $x = \theta_{i+1}{}^{\frac{1}{2}}, a = \theta_i^{\frac{1}{2}}, c = -\theta_i = -a^2, d = -2\text{arc\_length}/b$,then:

$$x^3 + ax^2 + cx + d = 0 \tag{3.9}$$

The solution:

$$x = \left(-a - \sqrt[3]{y_1} - \sqrt[3]{y_2}\right)/3 \tag{3.10}$$

where

$$\begin{aligned} y_1, y_2 = {} & \left(a^2 - 3c\right)a \\ & - 3\left((ac + 9d) \pm \sqrt{(ac - 9d)^2 - 4(a^2 - 3c)(c^2 - 3ad)}\right)/2 \end{aligned} \tag{3.11}$$

3.2.3 *Concentric Circle Path Mode*

3.2.3.1 How to Generate a Concentric Path

The concentric path has a series of concentric circles around the center of the workpiece. It is easy to implement by a polishing machine with a linear axis and a

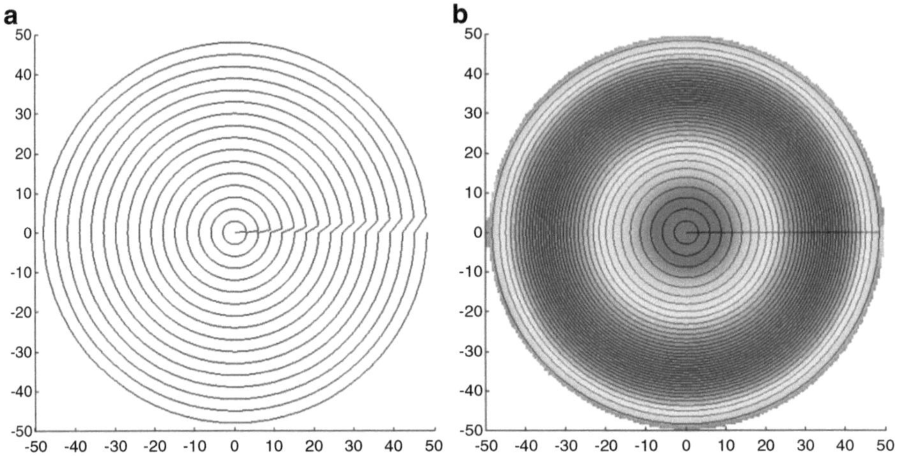

Fig. 3.8 Concentric circle path with (**a**) uniform pitch and (**b**) non-uniform pitch

rotating axis. If the path adopts the equal angle method, the generation rule can be expressed as:

$$\begin{cases} \rho_i = \text{floor}\big((i-1)/(360/\Delta\theta)\big) \cdot \Delta\rho \\ \theta_i = (i-1) \cdot \Delta\theta \end{cases} \tag{3.12}$$

where, i represents the path number, floor(x) means an integer, which is less than x, $\Delta\rho$ denotes the circle interval, and $\Delta\theta$ is the angle interval. An example of the equal angle concentric path is shown in Fig. 3.8a.

If we take the equal arc length method, the path can be expressed as Eq. (3.13):

$$\begin{cases} \rho_i = \text{floor}(\theta_i/360 + 1) \cdot \Delta\rho \\ \theta_i = \theta_{i-1} + 180/(2\pi) \cdot \text{arc\_length}/\rho_i \end{cases} \tag{3.13}$$

As a scanning path, concentric also has saltation (i.e. circle interval) along the scanning direction (i.e. radial direction). Thus, the circle interval should be confirmed carefully to avoid the appearance of grooves or ripples. In terms of uniformity, the equal arc length is preferable than the equal angle. The latter has more points in the central regions, and its edge path points are quite sparse.

3.2.3.2 Concentric Path with Variable Interval

If rotating symmetric errors exist on the surface form, then the path interval of a concentric circle path can be varied according to the distribution of the surface form. As shown in Fig. 3.8b, this path is similar to the spiral path with variable path interval. The tool can dwell in the higher region with longer time to remove more material.

Fig. 3.9 Some beautiful fractal pictures

3.2.4 *Various Fractal Path Modes*

3.2.4.1 Introduction to Fractal Art

Fractal theory is one of the main branches of non-linear science. It has been widely used in computer science, chemical biology, astronomy, geography and natural science, and economics, etc. A fractal is a mathematical set that has a fractal dimension that usually exceeds its topological dimension and may fall between the integers. Fractals are typically self-similar patterns, where self-similar means they are 'the same from near and from far'. Fractals may be exactly the same at every scale, or they may be nearly the same at different scales. The definition of fractal goes beyond self-similarity per se to exclude trivial self-similarity and include the idea of a detailed pattern repeating itself. Figure 3.9 gives some beautiful pictures with fractal pictures.

Fractal curves are one branch of fractal art, which can fill up a plane surface without self-crossing. The covering domain can be square or any other shape. Researchers have introduced fractal curves as paths used in deterministic optical surfacing systems. This subsection would introduce the frequently-used fractal curves and relating generation rules; readers can generate their fractal curves according to these instructions.

3.2.4.2 Common Fractal Curves

1. Koch Curve

In 1904, Swedish mathematician H. von Koch introduced his curve; because the curve segment is very like snow crystals, it is therefore also called the Snowflake Curve. Von Koch was not a famous mathematician at first, but his proposed curve is a typical fractal that is one of the most often studied and discussed. In the process, mathematicians have developed a system for the main method of drawing fractal,

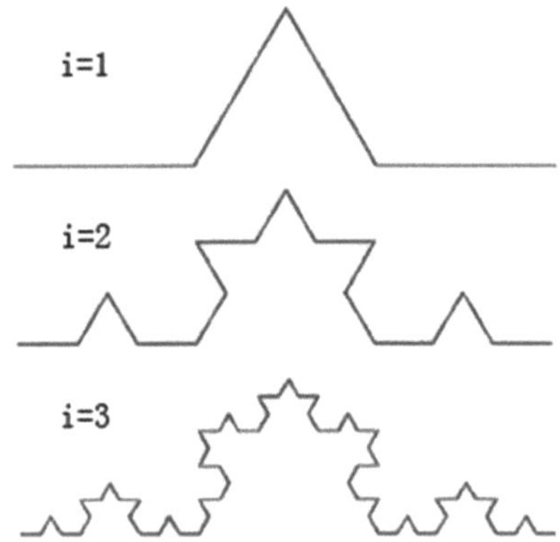

Fig. 3.10 The iterative generation process of Koch curve

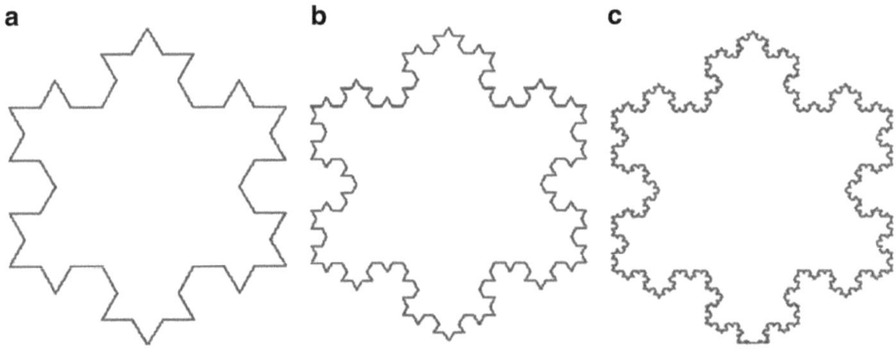

Fig. 3.11 Koch curves with (**a**) two iterations, (**b**) three iterations, and (**c**) four iterations

initiators and generators. The diversity of the fractal graphics start from here. You will see many curves similar to the Koch curve. The curve is composed of mutual joint line segments, and their complexity is like the nature of the coastline. Every fold also contains a more exquisite fold.

A sketch map of the iterative generation process is presented in Fig. 3.10. The generation rule is concluded as follows:

(a) Draw a line segment (assuming the length $L = 1$).
(b) The line segment is divided into three classes. Then, take the middle of the line as the bottom edge, making a trilateral shape, and then remove the lower side of the triangle, as shown in Fig. 3.10 ($i = 1$).
(c) Repeat to each line segment in the last step, as in Fig. 3.10 ($i = 2$).
(d) Repeat step (c) according to iteration times to get the curve.

If we take three sections of the Koch curve into a triangle synthesis, it can form beautiful snow crystals, like that shown in Fig. 3.11. Figures 3.11a–c have different iteration times from 2 to 4. This curve can describe a boundary such as the coastline of England; however, it has no power to fill a closed region uniformly.

Fig. 3.12 Peano curve generation process with multiple iterations

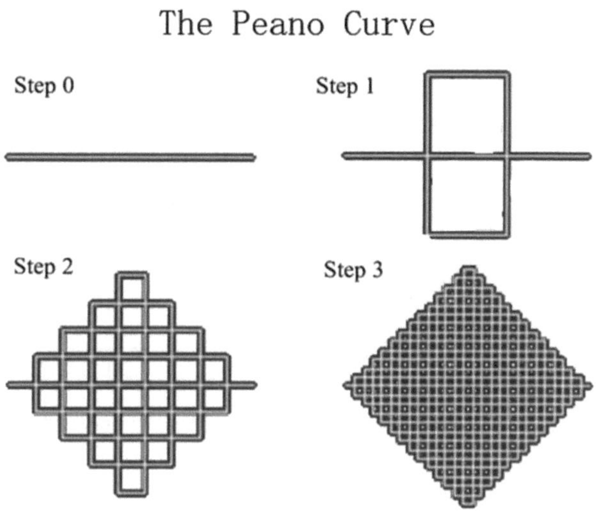

2. Peano Curve

In 1890, Giuseppe Peano found a curve to fill the plane, or he found something in the limited surface area curve that is the most efficient way of 'stretch'. In a biological point of view, if the organism must make effective use of limited space, they are likely to have taken such a structure. The production of the Peano curve is presented in Fig. 3.12 and is concluded as follows:

Step 0: draw a line segment as in Fig. 3.12 (Step 0).
Step 1: the line is divided into three classes, according to the Fig. 3.12; each line segment is combined with each other on the endpoint, instead of interlacing segmentation.
Step 2: step 1 is repeated for every segment generated in step 1.
Step 3: repeat step 2 as necessary.

The tool path is suitable for the optical surfacing process; it covers a square region with multiple line segments that are not self-crossing. There are numerous transformations from the Peano curve, as Fig. 3.13 shows: the left curve is Peano-1, the right is Peano-2 curve; they are so beautifully distributed in the space domain.

3. Hilbert Curve

In 1891, David Hilbert put forward a kind of curve that can fill the plane without any intersection; we called this the Hilbert curve. The curve attracted more attention of mathematicians than the Peano curve because it can travel every division unit of a plane in a non-interleaved way. This feature is also used to deal with the issue of image segmentation.

The Hilbert curve is a space-filling curve that visits every point in a square grid with a size of 2×2, 4×4, 8×8, 16×16, or any other power of 2. Applications of the Hilbert curve are in image processing, especially image compression and dithering. The basic elements of the Hilbert curves are 'cups' (a square with one open side, as shown in Fig. 3.14 [Step 0]) and 'joins' (a vector that joins two cups,

Fig. 3.13 Variations of Peano curves. (**a**) Peano-1 curve; (**b**) Peano-2 curve

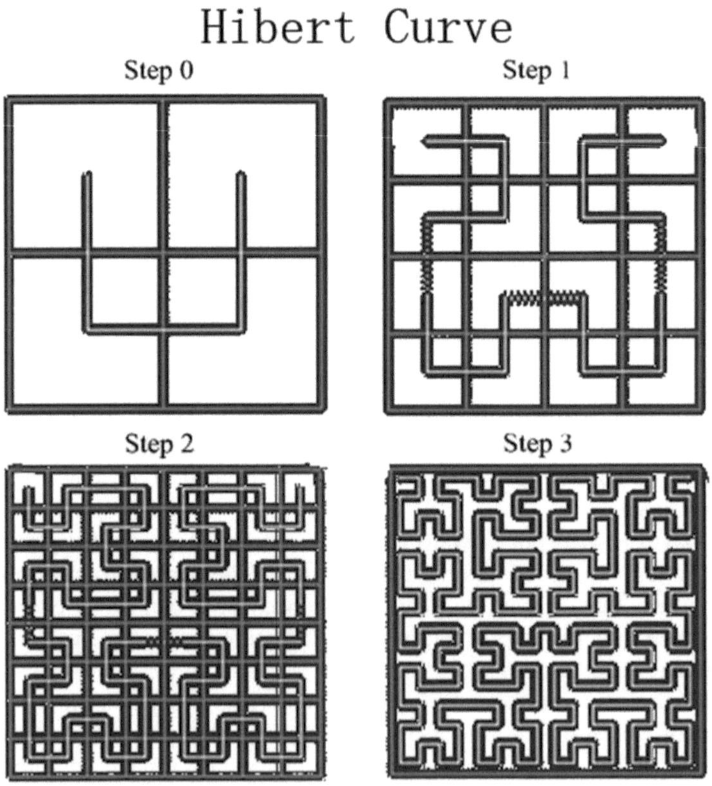

Fig. 3.14 Generation process of the Hilbert curve

Table 3.1 The symbol and meaning used in the L-System

Symbol	Meaning
F	Step forward for a unit length, and plot the line segment
G	Step forward for a unit length, but do not plot the line segment
+	Rotate in CCW direction with a specified angle
−	Rotate in CW direction with a specified angle

as shown by the dotted line in Fig. 3.14 [Step 1]). The 'open' side of a cup can be top, bottom, left, or right. As shown in Fig. 3.14, the first-order Hilbert curve is just a single cup that fills a 2×2 space grid. The second-order Hilbert curve replaces that cup by four (smaller) cups, which are linked together by three joins, as shown in Fig. 3.14 (Step 1). Every next order then repeats the process of replacing each cup by four smaller cups and three joins.

4. L-Systems for the Generation of Fractal Paths

Among lots of initiator and generator iteration methods, one of the most famous is the L-Systems proposed by American biologist Aristid Lindenmayer, also known as Parallel Rewriting Systems. In order to describe the process of plant growth and morphology, Lindenmayer developed a symbolic language set (formal language). The characteristic of this method is that it successfully modeled the iterative process of the line segment unit. At the same time, this method is also very easily achieved on computer simulation.

L-Systems, a method that mainly includes three parts: symbol definition (Alphabet), setting angle (Angle), the initial string (Axiom) with the rule string (Production Rules). Each are illustrated as follows:

Symbol definition. define L-Systems, a method by string in the process of writing; all need to use the character of symbols, including 'F', 'G', '+' and '−', as described in Table 3.1.

Angle setting. define the rotating angle with '+' and '−' symbols. The angle is usually between 2 and 50, representing (360/Angle). In terms of the Koch Curve, for example, its Angle = 6, on behalf of 60° rotating. In terms of the Peano and Hilbert curves, the Angle = 4.

Initial string. define the first string operation of the L-Systems method, equivalent to the starting unit of the 'starting and generating iterative method'. In terms of the Koch Curve, for example, the Axiom = F.

Rule string. defines the rewriting rules of each letter, equivalent to 'generated in the unit' of the 'starting and generating unit iterative method'. In terms of the Koch curve, for example, it is only one rule string, i.e. F = F + F − F + F.

So we can set Koch curve to L-Systems, Table 3.2 gives the generation process. In this format, we started with the initial string F in Step 0, then replaced every F with the rule string "F + F − F + F" in Steps 1–3. The L-Systems description of the Koch curve, Peano and Peano-2 curves, and the Hilbert curve are presented as

Table 3.2 Generation process of the Koch curve with the L-System

Step	Koch-curve symbols
0	F
1	F + F − F + F
2	F + F − F + F + F + F − F + F − F + F − F + F + F + F − F + F
3	……

Table 3.3 Generation rule of the Koch curve with the L-System

	Koch-curve symbols
Angle	6
Initial string	F
Rule string	F = F + F − F + F

Table 3.4 Generation rule of the Peano curve with the L-System

	Peano-curve symbols
Angle	4
Initial string	F
Rule string	F = F − F + F + F + F − F − F − F + F

Table 3.5 Generation rule of Peano-2 with the L-System

	Peano-2-curve symbols
Angle	4
Initial string	X
Rule string	X = XFYFX + F + YFXFY − F − XFYFX Y = YFXFY − F − XFYFX + F + YFXFY

Table 3.6 Generation rule of the Hilbert curve with the L-System

	Hilbert-curve symbols
Angle	4
Initial string	X
Rule string	X = −YF + XFX + FY− Y = +XF − YFY − FX+

Tables 3.3, 3.4, 3.5, and 3.6, respectively. With these generation rules, relative curves could be easily generated in computer simulation.

3.2.5 Surface Form-Based Paths

For a surfacing process in the rectangular coordinate, the tool can traverse a non-uniform polishing path based on the surface error map. It can gather more dwell points on the higher regions of the surface error map. Assuming the width of a square plane workpiece is 100 mm, the workpiece correlates with a data array $Z_{m \times n}$, the coordinate of a raster path is ($X_{s \times k}$, $Y_{s \times k}$). Then $X_{s \times k}$, $Y_{s \times k}$ can be denoted as Eq. (3.14).

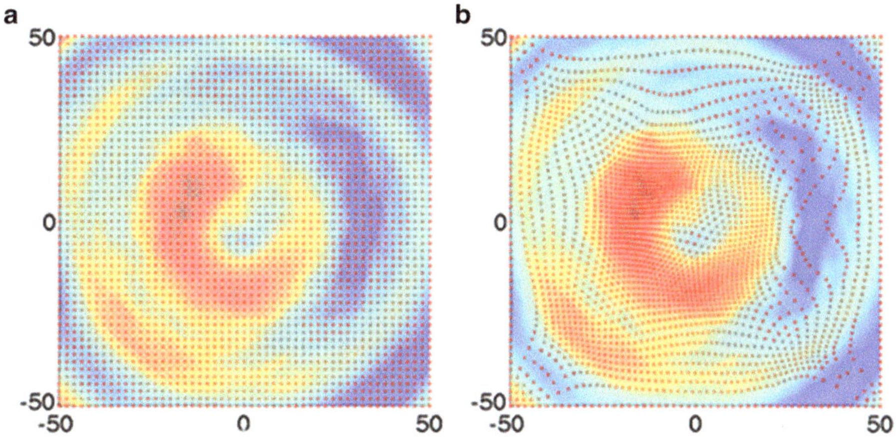

Fig. 3.15 Polishing path: (**a**) Raster path; (**b**) surface error-based non-uniform path

$$X = \begin{vmatrix} x_{11} & x_{12} & \cdots & x_{1k} \\ x_{21} & x_{22} & \cdots & x_{2k} \\ \cdots & \cdots & \cdots & \cdots \\ x_{s1} & x_{s2} & \cdots & x_{sk} \end{vmatrix} \quad Y = \begin{vmatrix} y_{11} & y_{12} & \cdots & y_{1k} \\ y_{21} & y_{22} & \cdots & y_{2k} \\ \cdots & \cdots & \cdots & \cdots \\ y_{s1} & y_{s2} & \cdots & y_{sk} \end{vmatrix} \tag{3.14}$$

We then define a weight factor p, as shown in Eq. (3.15), in which q is an adjusted coefficient (set as 5–10); a bigger q can increase the path density at a high region of surface form. p_1 is defined as Eq. (3.16).

$$p = 1 + q \cdot p_1 \tag{3.15}$$

$$p_1(i,j) = \frac{Z(i,j) - \min(Z)}{\max(Z) - \min(Z)} \tag{3.16}$$

The surface form-based path ($X_{s\times k}{}'$, $Y_{s\times k}{}'$) can be obtained from Eq. (3.17), where w is the iteration time.

$$\begin{cases} X'(i,j)^{w+1} = \dfrac{\sum\limits_{u=i-1}^{i+1}\sum\limits_{v=j-1}^{j+1} X(u,v)^w * p(u,v)^w}{\sum\limits_{u=i-1}^{i+1}\sum\limits_{v=j-1}^{j+1} p(u,v)^w} \\ X'(i,j)^{w+1} = \dfrac{\sum\limits_{u=i-1}^{i+1}\sum\limits_{v=j-1}^{j+1} Y(u,v)^w * p(u,v)^w}{\sum\limits_{u=i-1}^{i+1}\sum\limits_{v=j-1}^{j+1} p(u,v)^w} \end{cases} \tag{3.17}$$

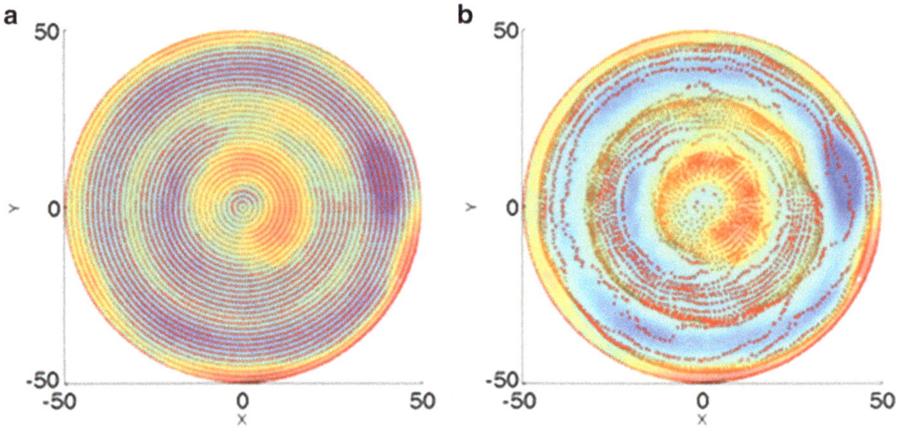

Fig. 3.16 Paths: (**a**) concentric circle path; (**b**) surface error-based non-uniform path

Set as $w = 100$, the non-uniform path derived from the raster path of Fig. 3.15a is given in Fig. 3.15b. This calculation took about 2 s in MATLAB. The colorized background represents the surface error map, and red points denote the path points. It is easy to see that the path density in the region of the lower surface form map is largely decreased. More dwell points are gathered on the high regions of the surface error map. If in the polar coordinate, for a concentric path shown in Fig. 3.16a, the revised non-uniform path is presented in Fig. 3.16b. The lower regions are generally avoided to be fabricated.

3.2.6 Boundary-Based Paths

With respect to workpieces with uncommon boundary, one common method for covering this region is truncating common paths by the boundary. For example, one can fill a hexagon and round square region with concentric circle or spiral paths, as shown in Fig. 3.17; red points are located in the inner region of the workpieces. When scanning such a path, in the edge region, the contact-type surfacing tool has to rise up to avoid risks, which is complex and time-consuming.

Another method is searching a path conforming to the boundary of the workpiece. For a workpiece with a known boundary, given as a vector of (ρ_b, θ_b), then the boundary-based path (ρ, θ) can be obtained according to Eq. (3.18):

$$\begin{cases} \rho_i = \text{ceil}(\theta_i/2\pi)/n \cdot \text{interp}(\theta_b, \rho_b, \theta_i) \\ \theta_i = \theta_{i-1} + d\theta \end{cases} \tag{3.18}$$

where ceil($\theta_i/2\pi$) indicates the minimal integer which is larger than $\theta_i/2\pi$. n represents the number of turn, interp denotes the one-dimension interpolation, and $d\theta$ represents the angle interval between adjacent points.

If the boundary is unknown, then it can be measured by a coordinate measurement machine (CMM). As an example, the hexagon workpiece was filled with the boundary-

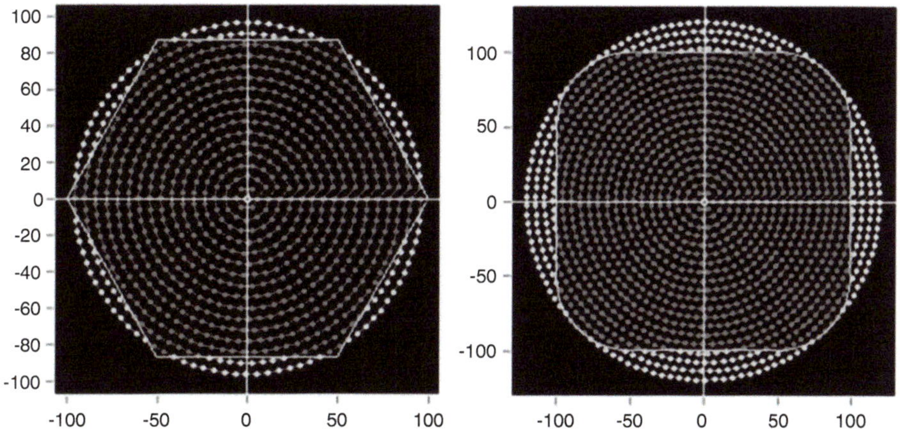

Fig. 3.17 Truncating path for a hexagon region and round square

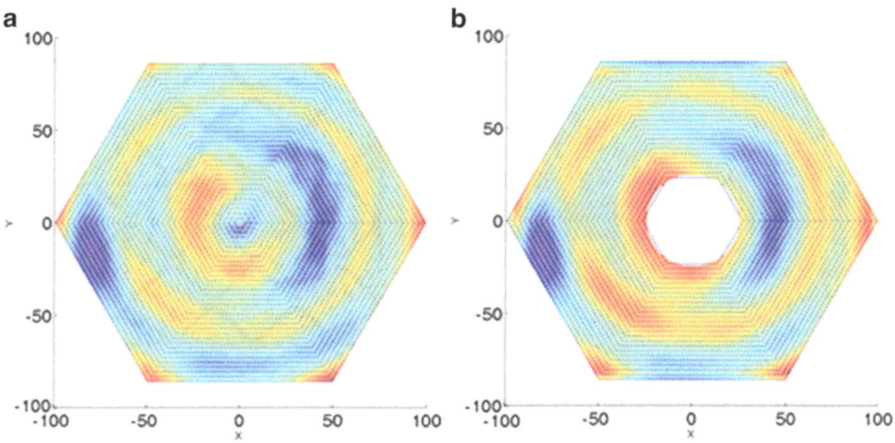

Fig. 3.18 Boundary-based path for a hexagon region (**a**) without and (**b**) with a central hole

based path shown in Fig. 3.18. It has equal arc length. Its subline is similar to its boundary, so the surfacing tool can finish the workpiece without rising and slumping.

3.3 Peano-Like Tool Path

3.3.1 Insufficiencies of the Peano Path

The Peano path is a type of fractal path (as shown in Fig. 3.19). The path density increases with the iteration times. The Peano path contains nine straight line segments in the first iteration (Fig. 3.19a). In the second iteration (Fig. 3.19b), each line segment in the first iteration is replaced by nine segments. Thus, the path contains 81 segments.

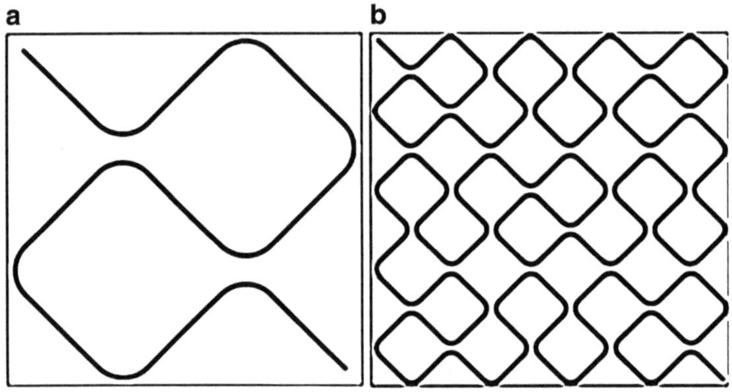

Fig. 3.19 Peano path with (**a**) one iteration and (**b**) two iterations

Fig. 3.20 Peano-like path covering a square region

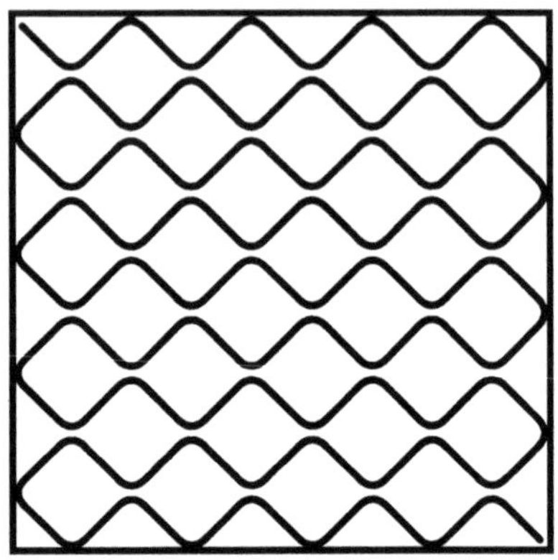

This path generation method is rather inflexible. The number of path segments is 9^n, where n is the nth iteration. The region covered is always square. The path generation may be revised, as illustrated in Fig. 3.20, to be called the Peano-like path. Here, one zigzags between two opposite boundaries. Within the same square region, the finished path is the same for both generation methods. With Peano-like paths, the number of path segments to be used to fill up a region can be rather flexible and the region needs not be square.

3.3.2 Peano-Like Path Construction on Aspherical Surfaces

An aspheric surface can be represented as Eq. (3.19):

$$S = \begin{bmatrix} x \\ y \\ z \end{bmatrix} = \begin{bmatrix} r\cos\theta \\ r\sin\theta \\ z(r) \end{bmatrix} \tag{3.19}$$

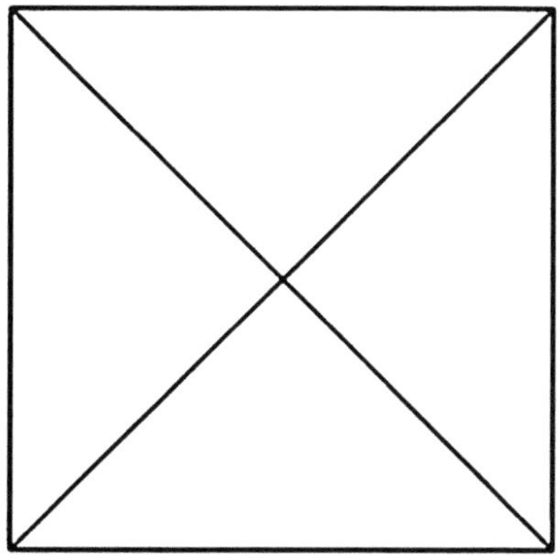

Fig. 3.21 Path and coverage on a square region

in which the Z-axis is aligned to the axis of revolution, and r is the distance of a surface location from the Z-axis. The form of the surface is given by $z(r)$. The surface is parameterized by variables u and w, with $r = r(u)$ and $\theta = c + (d - c)w$. The first derivatives of S with respect to u and w are expressed as Eq. (3.20):

$$S_u = \begin{bmatrix} \cos\theta \\ \sin\theta \\ z_r \end{bmatrix} r_u \qquad S_w = \begin{bmatrix} -\sin\theta \\ \cos\theta \\ 0 \end{bmatrix} r\theta_w \tag{3.20}$$

and the corresponding norms are shown as Eq. (3.21):

$$|S_u| = |r_u|\sqrt{1 + z_r^2} \quad |S_w| = r|d - c| \tag{3.21}$$

Thus, $|S_w|$ is independent of θ and w. In addition,

$$S_u \cdot S_w = \begin{bmatrix} \cos\theta \\ \sin\theta \\ z_r \end{bmatrix} r_u \cdot \begin{bmatrix} -\sin\theta \\ \cos\theta \\ 0 \end{bmatrix} r\theta_w = 0 \tag{3.22}$$

which implies that S_u and S_w are perpendicular to each other at all locations on the surface.

The coverage rate C is defined as the ratio between the path length in a region and the area of the region. For a square region $p \times p$ (Fig. 3.21), a path segment can follow a diagonal line either joining the lower left and upper right corners, or joining the upper left and lower right corners of the region. In this case, one has:

$$C = \frac{\text{length}}{\text{area}} = \frac{\sqrt{2}p}{p^2} = \sqrt{2}/p \tag{3.23}$$

Suppose a surface is divided into small surface regions. If there is a tool path on the surface such that the coverage rate is the same for all the surface regions, the surface is considered to be uniformly covered by the tool path.

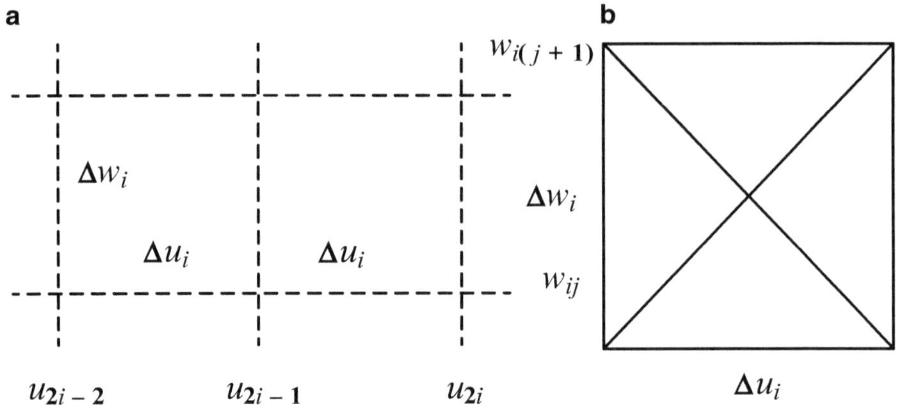

Fig. 3.22 Path on the *u*-*w* domain: (**a**) partitioning of the domain; (**b**) path segments within a partition

Suppose a tool performs a certain processing action on a surface as it follows a tool path on the surface. Let the processing rate P be defined as the amount of task performed as the tool passes through a surface region divided by the area of the surface region. For instance:

$$P = \frac{\text{time} \times \text{task rate}}{\text{area}} = \frac{\frac{\text{length}}{\text{feed rate}} \times \text{task rate}}{\text{area}} = C \times \frac{\text{task rate}}{\text{feed rate}} \tag{3.24}$$

For the case of polishing, the task rate may be taken as the amount of material removed from the surface by the polishing tool in one time unit. A surface is considered uniformly processed if P is the same at all locations on the surface. Suppose the task rate and the feed rate are constant, P is uniform over the surface if C is uniform over the surface.

The domain of an aspheric surface $S(u, w)$ is divided into u-strips by u-iso-parametric lines (Fig. 3.22a). The ith pair of u-strips are in the range of $u_{2i-2} < u < u_{2i-1}$ and $u_{2i-1} < u < u_{2i}$, where $u_0 = 0$. They are further partitioned along the w-direction at w_{i0}, w_{i1}, w_{i2} $\cdots$ such that $w_{i0} = 0$. All rectangular partitions within the ith pair of u-strips have the same size. The width and height of these partitions are Δu_i and Δw_i, respectively, where $\Delta u_i = u_{2i-1} - u_{2i-2} = u_{2i} - u_{2i-1}$ and $\Delta w_i = w_{i(j+1)} - w_{ij}$.

These u- and w- iso-lines are perpendicular to each other in the Cartesian space. Moreover, $|S_w|$ is constant for a given u and does not depend on w. The surface patch in the Cartesian space corresponding to the region $\Delta u_i \times \Delta w_i$ may be approximated by a square region $p \times p$ if:

$$\begin{cases} \Delta u_i = p/|S_u| \\ \Delta w_i = p/|S_w| \end{cases} \tag{3.25}$$

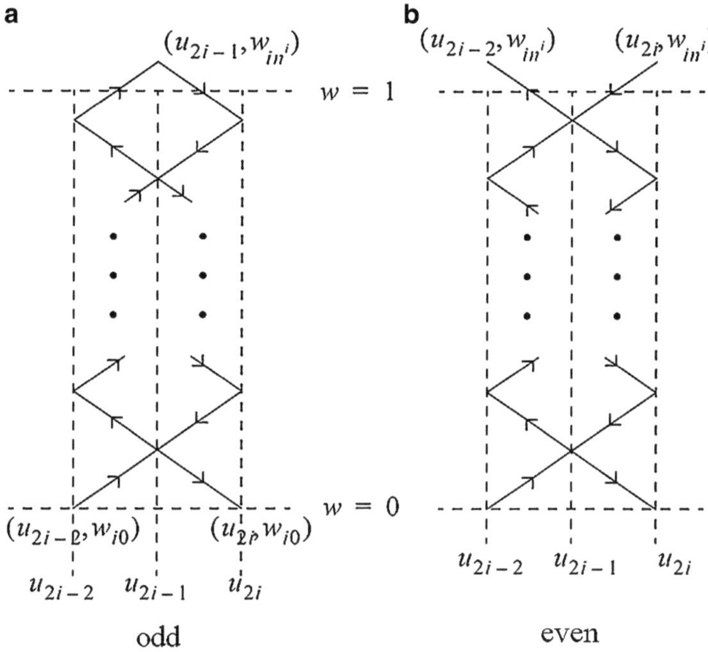

Fig. 3.23 Path on a pair of u-strips. (**a**) n^i odd; (**b**) n^i even

where the partial derivatives are evaluated at u_{2i-1} and p is small relative to the local radius of curvature of the surface. In this case, the area of a surface patch is approximately equal to p^2. Suppose there is a path segment in each of the rectangular partitions, and the path segment is along either one of the two diagonal lines (Fig. 3.22b). The length of a diagonal path segment is $\sqrt{2}p$ and the coverage rate is simply $= \sqrt{2}/p$.

According to Eqs. (3.23) and (3.24), one may use $p = \frac{\sqrt{2}}{P} \times \frac{\text{task rate}}{\text{feed rate}}$ for the calculation of Δu_i and Δw_i. For uniform processing, P is a constant. The feed rate is typically constant along the tool path. If the part surface has gentle curvatures compared with those of the polishing tool, the task rate may be assumed constant throughout the surface. In such a case, one may use a constant p for path generation for the whole surface.

3.3.2.1 Scanning the u-Strips

Paths can be generated by scanning up and down the u-strips. When one pair of u-strips is filled, the process continues with the next pair of u-strips.

Upward scanning in the ith pair of u-strips starts at (u_{2i-2}, w_{i0}), and the path zig-zags between u_{2i-2} and u_{2i-1}, while the w-coordinate is increased by Δw_i from w_{ij} to $w_{i(j+1)}$ at each step upward. Upward scanning stops when $w_{in^i} > 1$.

If n^i is odd (Fig. 3.23a), set $J^i = n^i$, otherwise (Fig. 3.23b) set $J^i = n^i - 1$. This ensures the uppermost point is on the u_{2i-1} iso-line.

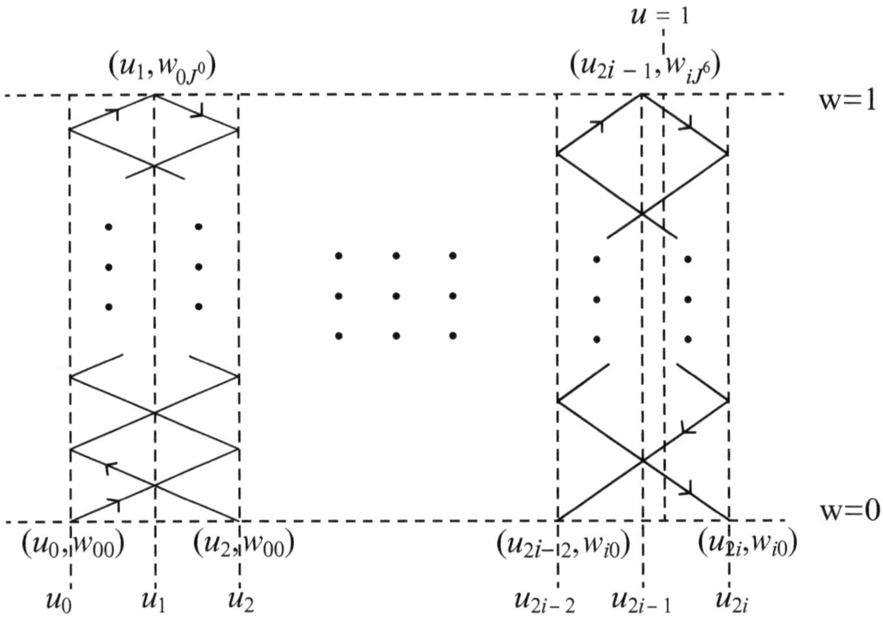

Fig. 3.24 Termination of u-direction increment

Downward scanning starts at (u_{2i-1}, w_{iJ^i}) and the path zig-zags between u_{2i-1} and u_{2i}, while the w-coordinate is decreased by Δw_i from $w_{i(j+1)}$ to w_{ij} at each step downward. Scanning stops at (u_{2i}, w_{i0}). Subsequently, path points are scaled to fill up the range of w between 0 and 1: $w_{ij} = w_{ij}/w_{iJ^i}$ for $J = 1, \cdots, J^i$.

Resulting path points in the ith pair of u-strips are in the order of:

$$\left\{(u_{2i-2}, w_{i0}), (u_{2i-1}, w_{i1}), (u_{2i-2}, w_{i2}), \cdots, (u_{2i-1}, w_{iJ^i}), \left(u_{2i}, w_{i(J^i-1)}\right), \cdots, (u_{2i}, w_{i0})\right\}$$

with $w_{iJ^i} = 1$.

Scanning of the next pair of u-strips is executed only if $u_{2i-1} < 1$. There is no further u-direction increment after the current scan if $u_{2i} > 1$ (Fig. 3.24). Upon termination of scanning, set $I = i$. The path points are scaled to fill up the range of u between 0 and 1: $u_i = u_i/u_{2I}$ for $i = 1, \ldots, 2I$.

The tool path is obtained from piecing together path points of the successive scanning of u-strips. Thus, the resulting path contains path points in the order of:

$$\{(u_0, w_{00}), \ldots, (u_1, w_{0J^0}), \ldots, (u_2, w_{00}), \ldots, (u_{2I-1}, w_{I0}), \ldots, (u_{2I-1}, w_{IJ^I}), \ldots, (u_{2I}, w_{J0})\}$$

and $u_{2I} = 1$.

The proposed Peano-like path was applied to generate a path for an off-axis spherical segment. The form of the surface was given by $z(r) = \sqrt{R^2 - r^2}$, with a

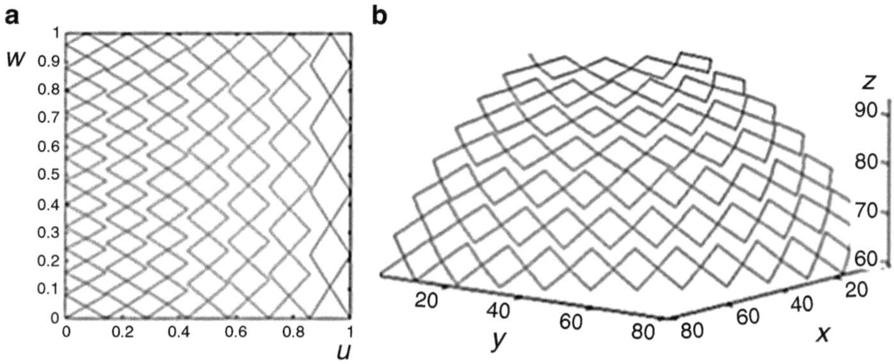

Fig. 3.25 Peano-like path on an off-axis spherical segment. (**a**) *u*-*w* domain; (**b**) Cartesian domain

radius of $R = 100$ mm and $r(u) = R \cos(36.87° + 38.13° \times u)$. Path pitch p was set at 5 mm.

The tool path was one continuous path, as shown in Fig. 3.25. The path started and ended at $(u, w) = (0, 0)$ and $(1, 0)$, respectively (Fig. 3.25a). These corresponded to the lower left and top left corners of the surface segment (Fig. 3.25b). Path segments within each pair of *u*-strips form diamonds of the same size (Fig. 3.25a). The *u*-strips correspond to circular strips on the surface segment (Fig. 3.25b). The diamonds on the *u*-strips correspond to the square-like elements on the circular strips.

Path segments were diagonal lines of the $\Delta u \times \Delta w$ regions in the *u*-*w* domain. The length of the four edges of the surface patch corresponding to a $\Delta u \times \Delta w$ region was approximately equal to p. The surface patch was approximately square and the length of the path segment was approximately $\sqrt{2}p$, as the path pitch was small compared with the local radii of curvature of the surface. Thus, each path segment was approximately orthogonal to its adjacent path segments.

The number of path segments in a *u*-strip depended on the curvatures and the length of *w*-iso-lines bordering the *u*-strip. Thus, there were more path segments in a *u*-strip near the bottom of the surface compared with that near the top.

In the current surface, not only were Δw values all the same within each *u*-strip, but the width of all *u*-strips (Δu) were also the same as the surface and had constant curvature along *u*-iso-lines.

The proposed path was expected to have better uniformity of coverage compared with scanning paths. All path segments of the proposed path had similar length, and the segments and turns were evenly distributed over the inner surface as well as near the edges of the surface. In contrast, for scanning paths (Fig. 3.26), turns were at the edges of the surface only, and path segments were long in the inner surface and short near the edges.

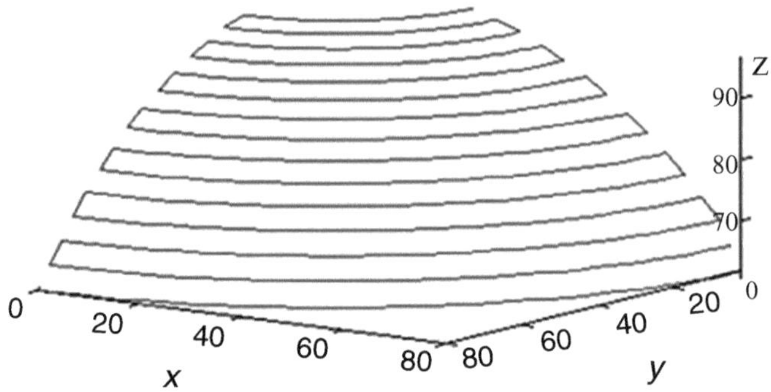

Fig. 3.26 Scan path on an off-axis spherical segment

3.3.2.2 Paths on Aspherics with Full Rotational Symmetry

There are four edges in the off-axis surface in Fig. 3.25. When an aspheric surface has full rotational symmetry, the surface has, at most, two edges. A case illustration is given in Fig. 3.27a in which a path was generated for a spherical surface with the same surface form and radius as in Fig. 3.25 but spanned a range of 360°. While uniform coverage of the surface was maintained in the same style as in Fig. 3.25, rotational symmetry of the path was not present where the left and right edges met (Fig. 3.27a).

To further enhance the symmetry of the tool path, a return path from the last point of the last u-strip back to the start point of the first u-strip is added as follows (Fig. 3.28):

$$\{(u_{2I}, w_{I0}), (u_{2I-1}, -w_{I0}), (u_{2I-2}, w_{I0}), \ldots, (u_2, w_{00}), (u_1, -w_{01}), (u_0, w_{00})\}$$

Subsequently, w-coordinates of all path points are further scaled so that all points lie within a unity range in the w-direction:

$$w_{ij} = \frac{J^i}{1+J^i} w_{ij} \text{ for } i = 0, \ldots, I \text{ and } j = 1, \ldots, J^i.$$

A path that includes the return path for the surface was generated, as shown in Fig. 3.27b-i. It can be seen that, with the return path, both uniform coverage and rotational symmetry of the tool path are maintained. The u-w domain of the path is shown in Fig. 3.27b-ii. Although the w-coordinate of some of the path points were below 0, the range of w-coordinate of all points on the same w-iso-line was kept within unity.

The same type of uniform coverage and rotational symmetry can be seen in Fig. 3.29, in which the method was applied to a parabolic surface. The surface had the form $z(r) = 200 - 0.005r^2$ and $r(u) = 80 - 30u$. It had the same base diameter as the surface in Fig. 3.28. The path pitch was set to 3 mm.

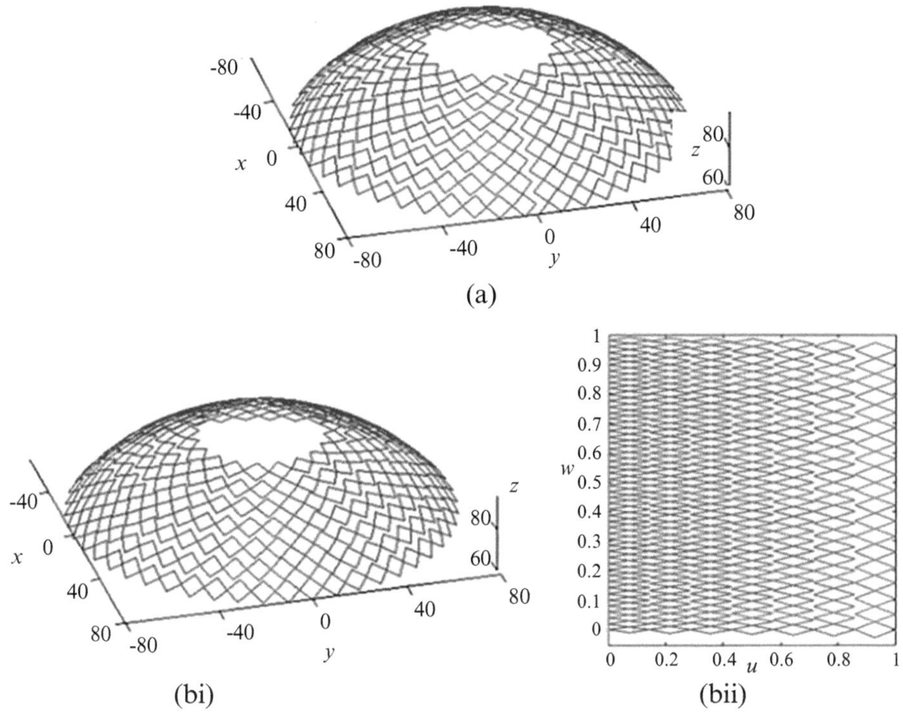

Fig. 3.27 Peano-like paths on a spherical surface with full rotational symmetry: (**a**) without return path; (**b-i**) with return path, Cartesian domain; (**b-ii**) with return path, *u*-*w* domain

3.3.2.3 Paths on Aspherics with One Edge

When an aspherical surface with full rotational symmetry is closed on top, the surface is left with only one edge. The same parabolic surface in Fig. 3.29 was considered for $r < 50$ mm. The same tool path method was applied and the path obtained is shown in Fig. 3.30. The surface is closed on top. One can see clustering of path segments near the vertex of the surface. Away from the vertex, each path segment is approximately orthogonal to its adjacent segments. As one moves toward the vertex, path segments seem to point more and more toward the vertex.

This clustering of paths segments can be related to the surface parameterization. Away from the vertex, the surface patch that corresponds to a $\Delta u \times \Delta w$ region is approximately square. The shape is slightly distorted and becomes somewhat trapezoidal for surface patches close to the vertex.

In the evaluation of Δw_i (Eq. (3.25)), partial derivatives of the surface are evaluated along the u_{2i-1} iso-lines. Square approximation of the shape of the surface patches may result in slight under-estimation of the area and the length of

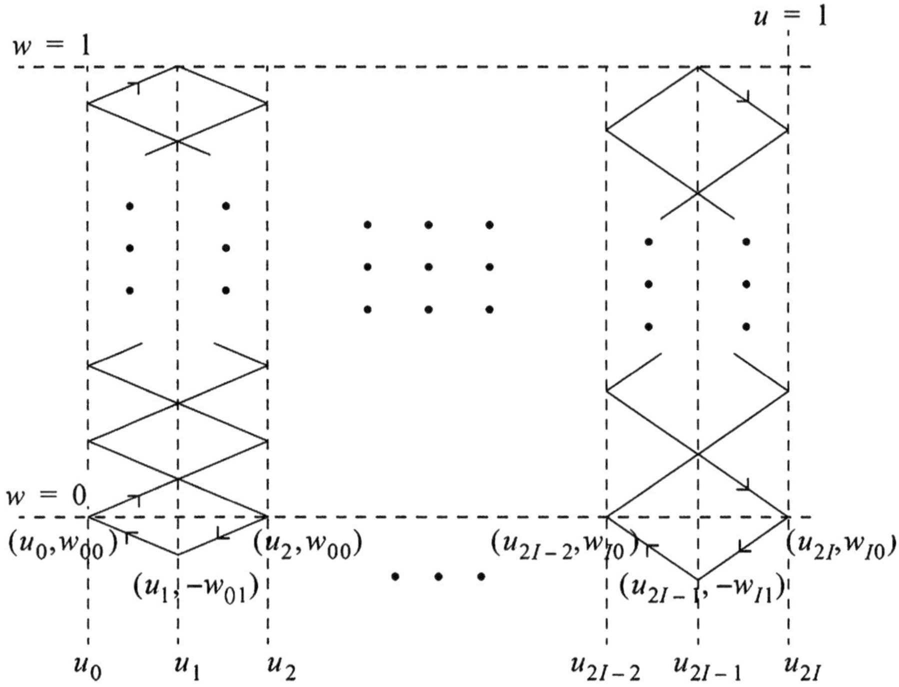

Fig. 3.28 Including a return path from (u_{21},w_{10}) back to (u_0,w_{00})

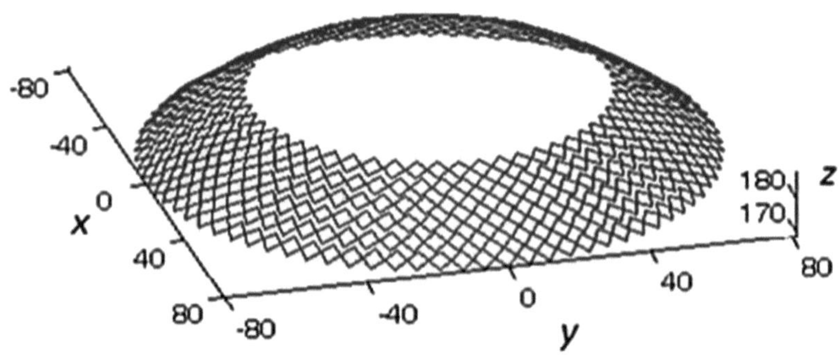

Fig. 3.29 Peano-like path on a parabolic surface

the diagonal line for the surface patch on one side of the iso-line and slight over-estimation of the corresponding entities on the other side. Away from the vertex, the overall effect of the surface parameterization on the uniformity of coverage may not be significant. Yet, the alignment of path segments toward the vertex for surface patches very close to the vertex may still cause non-uniformity of coverage near the vertex.

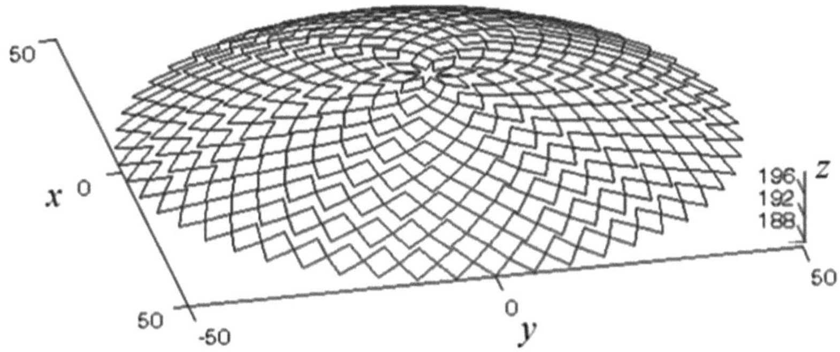

Fig. 3.30 Peano-like path on a parabolic surface with one edge, $r < 50$ mm

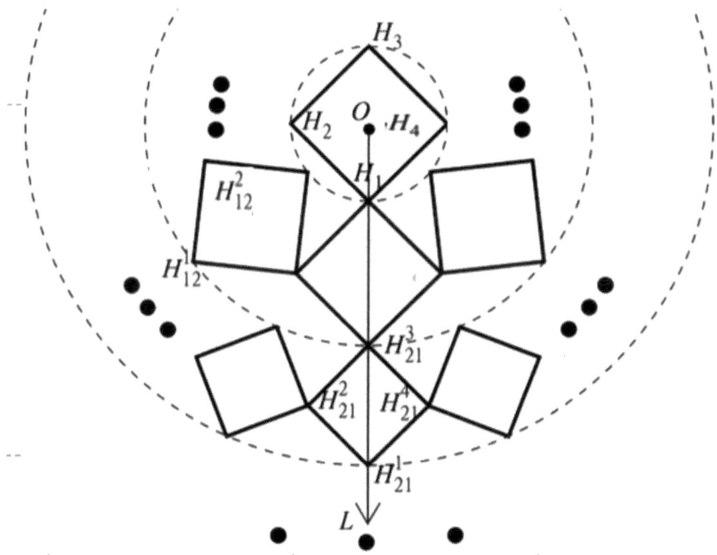

Fig. 3.31 Circular region covered by square elements

3.3.2.4 Square Elements in a Circular Region

Consider a pattern consisting of square elements arranged within a circular region on a plane (Fig. 3.31). The region is divided into a central circular region and N concentric annular regions. The central region contains one square element and the corners of the square touches the boundary of the central region. The ith annular region contains n_i square elements of the same size. They are symmetrically arranged so that each corner of a square touches either a corner of an adjacent square or a boundary of the region.

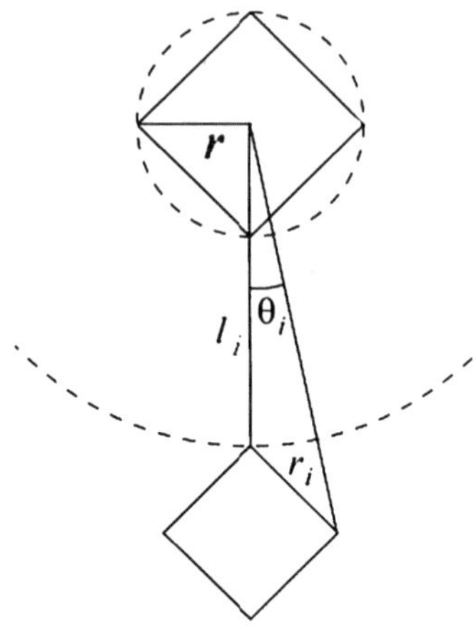

Fig. 3.32 Geometric entities with a circular region

The path length within the central region is $L_c = 4\sqrt{2}\Delta r$ and the area of the region is $A_c = \pi r^2$, where r is the radius of the central region (Fig. 3.32). The coverage rate for the central region is $C_c = L_c/A_c$. Let r_i be the length of a side of a square element in the ith annular region (Fig. 3.32). According to sine law,

$$r_i = \frac{\sin \theta_i}{\sin (45^\circ - \theta_i)} \Delta l_i \tag{3.26}$$

where $\theta_i = 180^\circ/n_i$ and l_i are the inner radius of the ith annular region. One has:

$$\begin{cases} l_1 = r \\ l_i = l_{i-1} + \sqrt{2} r_{i-1} \quad \text{for} \quad i > 1 \end{cases} \tag{3.27}$$

Using Eqs. (3.26) and (3.27), one can recursively calculate the dimensions of the square elements and the annular regions in terms of r, provided the n_i's are specified for the annular regions.

For the ith annular region, the path length is $L_i = 4n_i r_i$, the area $A_i = \pi(l_{i+1}^2 - l_i^2)$, and the coverage rate $C_i = L_i/A_i$. For the case of a circular region with a central region and N annular regions, the radius of the entire circular region is $r_c = l_N + \sqrt{2} r_N$.

The number n_i affects the coverage rate in each annular region. When the outer radius of the (i–1)th annular region is fixed, as n_i is increased, the size of the square elements in the ith annular region is reduced. This leads to an increase in the path length, decrease in the area of the region, and increase in the coverage rate. One can select n_i so that the coverage rate C_i is closest to that of the central region. This can be done by systematically increasing the number of square elements in an annular region and calculating the corresponding coverage rate. The initial guess of n_1 is 5, as from geometry, $n_1 = 4$ implies $r_1 \to \infty$. After one has determined n_1, the initial guess of n_i can be set at n_{i-1} for $i > 1$.

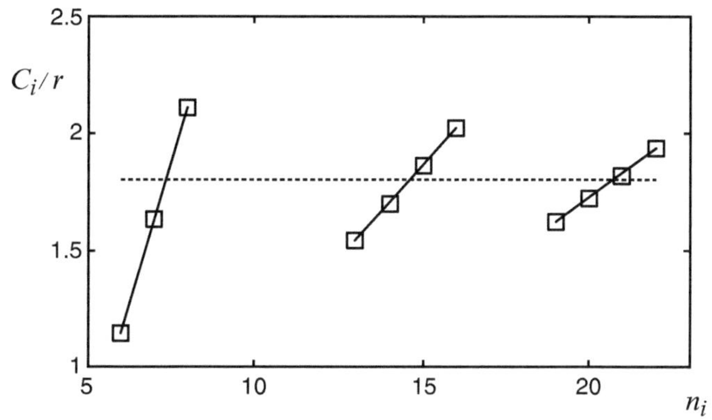

Fig. 3.33 Coverage rate C_i increases with n_i

Table 3.7 Coverage rate and entities of the circular region with optimized n_i's

i	n_i	r_i/r	C_i/r	r_c/r
c	NA	NA	1.801	1
1	7	1.314	1.634	2.858
2	15	1.091	1.861	4.402
3	21	1.105	1.824	5.965
4	27	1.116	1.800	7.543
5	33	1.126	1.782	9.133

Some of the calculation results are shown in Fig. 3.33. The coverage rate of the central region C_c is indicated by the horizontal dashed line. The calculations confirm that the coverage rate increases with n_i. One can make C_i close to C_c by a proper choice of n_i. Moreover, the values are getting closer as i increases.

The finally selected n_i's and the corresponding size of the square elements and the coverage rates are tabulated in Table 3.7, up to $i = 5$. It can be seen that the coverage rate in the annular regions are rather close to that in the central region. The largest discrepancy of about 9 % is in the first annular region and the discrepancy is about 3 % or below in the other regions. Also, the relative size of the square elements r_i/r becomes rather stable for $i \geq 2$.

3.3.2.5 Paths on the Vertex Cap

Shown in Fig. 3.34 is the part of an aspherical surface near the vertex. The vertex cap of radius r_c refers to a small region of the surface near the vertex where $r \leq r_c$.

Consider a tool path on the vertex cap such that its orthogonal projection onto the base has the shape of the path pattern, as shown in Fig. 3.31. Both the path length and the surface area of the vertex cap are approximated by the path length and the area of the base.

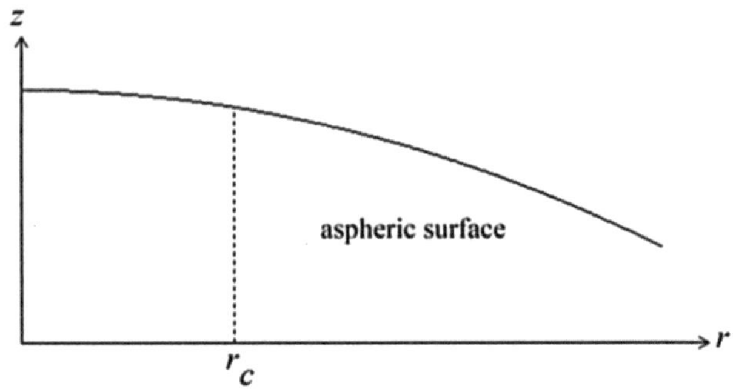

Fig. 3.34 Vertex cap of an aspherical surface

Table 3.8 Radius of the central region of the vertex cap for various N

N	r/p
c	1.274
1	1.170
2	1.254
3	1.270
4	1.271
5	1.268

For the case of a base with one central region and N annular regions, the total area of the base region is πr_{c}^2 and the total path length is $L_{\mathrm{c}} + \sum_{i=1}^{N} L_i$. Assuming the coverage rate over the vertex cap is $\sqrt{2}/p$, as for the rest of the surface, one has $\frac{\sqrt{2}}{p} = \frac{1}{\pi r_{\mathrm{c}}^2}\left(L_{\mathrm{c}} + \sum_{i=1}^{N} L_i\right)$ from which one obtains r/p for various values of N, which are tabulated in Table 3.8. It seems the ratio r/p is almost constant except for $N = 1$.

For aspherical surfaces with one edge, the Peano-like path can be used for $r > r_{\mathrm{c}}$, and the above path pattern can be used for $r < r_{\mathrm{c}}$.

Suppose the four corners of the square in the central region are designated by H_1 through H_4 (Fig. 3.31). Point H_{ij}^k represents the kth corner of the jth square element in the ith annular region.

Points on the planar surface in the figure are located by their polar coordinates with pole O and polar axis L. The polar axis passes through the first square element in each of the annular regions. The coordinates of H_1 through H_4 are (r, 0°), (r, 270°) $\cdots$ (r, 90°), respectively. Likewise, from geometry as indicated in Fig. 3.31, the coordinates of H_{ij}^1 through H_{ij}^4 are:

$$\left(l_i + \sqrt{2}r_i, (n_i + 1 - j) \times \frac{360°}{n_i}\right), \left(\sqrt{l_i^2 + \sqrt{2}l_i r_i + r_i^2}, (n_i + 1/2 - j) \times \frac{360°}{n_i}\right),$$
$$\left(l_i, (n_i + 1 - j) \times \frac{360°}{n_i}\right) \text{ and } \left(\sqrt{l_i^2 + \sqrt{2}l_i r_i + r_i^2}, (n_i + 3/2 - j) \times \frac{360°}{n_i}\right)$$

respectively.

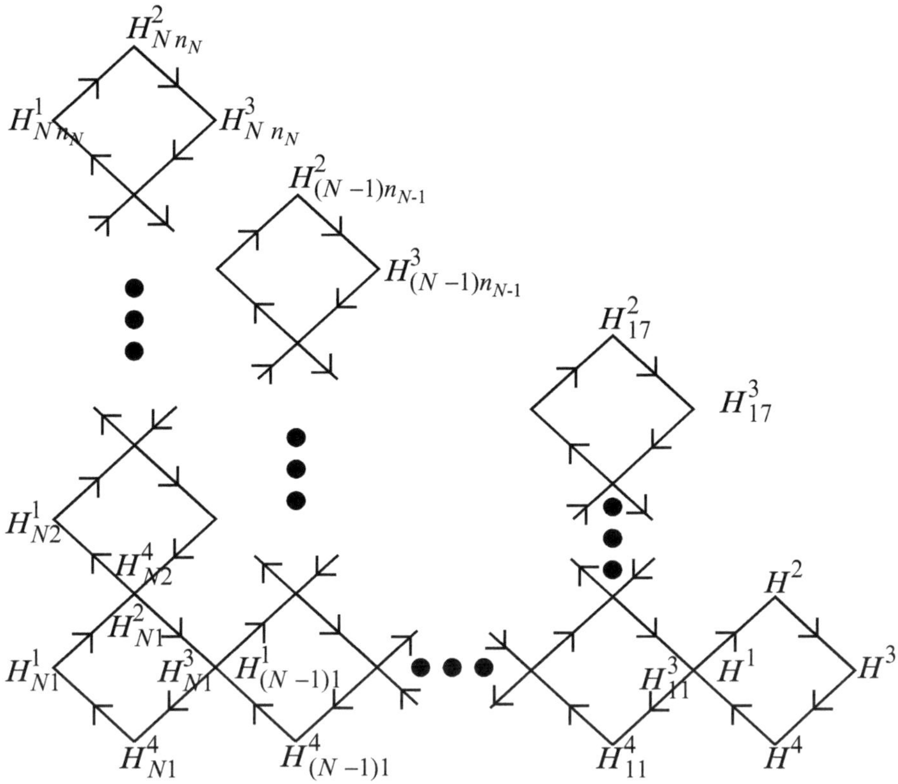

Fig. 3.35 Directed path segments inside a circular region

The square elements in the circular region can be mapped to the pattern in Fig. 3.35, while recognizing H^4_{i1} and $H^2_{in_i}$ denote the same location. Directions are added to the line segments to show that a path can be constructed to go through these square elements in a way similar to the path going through all path segments in Fig. 3.28.

One can follow the path points up and then down the square elements of the first column in Fig. 3.35 in order of $H^1_{N1}H^2_{N1}H^1_{N2}\cdots H^1_{Nn_N}H^2_{Nn_N}H^3_{Nn_N}H^4_{Nn_N}\cdots H^4_{N2}$. Then one follows the second column as $H^1_{(N-1)1}H^2_{(N-1)1}\cdots H^3_{(N-1)2}H^4_{(N-1)2}$ and then the remaining columns in order. After reaching H^4_{12} of the second to last column, one passes the four points $H^1H^2H^3H^4$ in the last column and then follows the return path $H^3_{11}H^4_{11}H^3_{21}\cdots H^3_{N1}H^4_{N1}$.

A top view of tool paths in the vicinity of the vertex cap of a parabolic surface are presented in Fig. 3.36. The tool path in Fig. 3.36a is the top view of the tool path in Fig. 3.30. The Peano-like path was used. One can clearly see the effects of surface parameterization on the tool path near the vertex. Far away from the vertex, path segments from each pair of u-strips formed a string

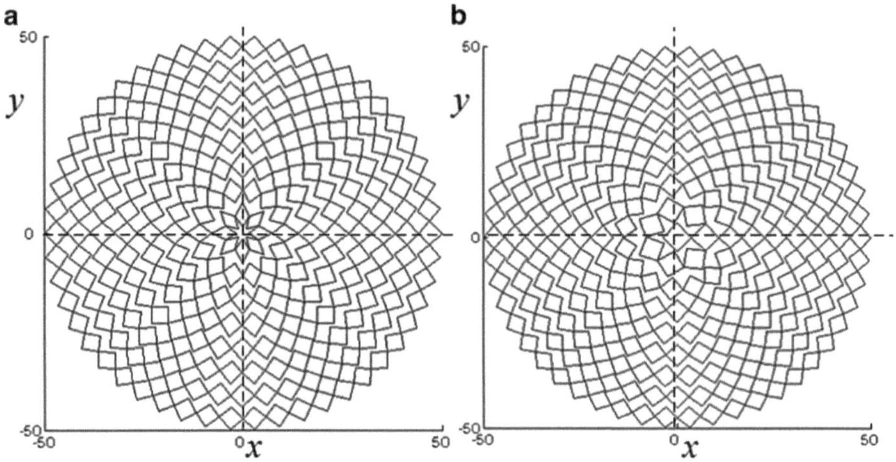

Fig. 3.36 Peano-like path on the parabolic surface near the vertex: (**a**) without and (**b**) with the vertex cap/square elements

of square-like elements, after accounting for distortion due to orthogonal projection. Closer to the vertex, the elements had the shape of kites, although the local slope was small and the distortion due to projection was negligible. Very close to the vertex, path segments were clustered together and aligned toward the vertex.

For the path in Fig. 3.36b, square elements were adopted within the vertex cap and the Peano-like path was adopted outside the vertex cap. The same path pitch of 3 mm was used as in Fig. 3.36a. The vertex cap was chosen to contain one central region and two annular regions. According to Tables 3.7 and 3.8, the radius of the top view of the vertex cap was $r_c = 16.56$ mm. The surface profile in Fig. 3.34 corresponded to the current parabolic surface and r_c was as marked in the figure. It can be seen that under the current circumstance of a small vertex cap, planar approximation is reasonable for the calculation of the local tool path.

Inside the vertex cap, there is one square element in the centre, instead of clustered path segments. Within the annular regions, path segments form square elements which are like the square-like elements on the surface far from the vertex. The size of the square elements was adjusted and uniformity of coverage was maintained, as suggested in Table 3.7. In particular, there is no concentration of path segments near the vertex. It seems there is not much distortion of the square-like elements beyond the second annular region, although they are computed based on the Peano-like path in the previous section. According to calculation, the size of the square elements in the second annular region and those outside the vertex cap differ by about 2 % only (Fig. 3.37).

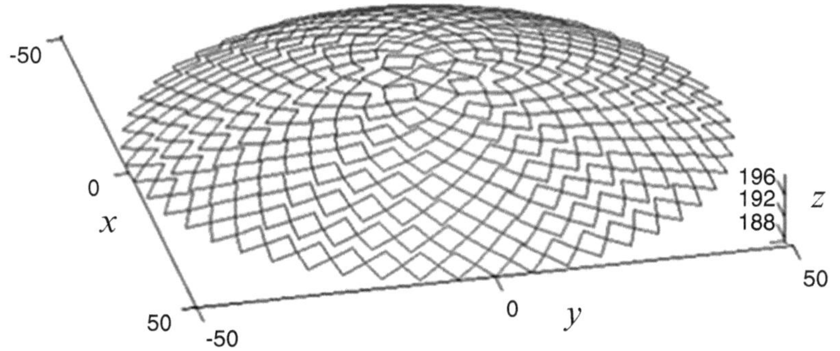

Fig. 3.37 Peano-like path on parabolic surface near the vertex cap, 3D view. *3D* three dimensional

3.4 The Influence of Paths on the Convergence Rate of Surface Form

3.4.1 Polishing Tool Paths

Polishing always pursues a high convergence rate of surface form, as well as best surface texture. Polishing paths have been confirmed as the most important factor influencing the convergence rate and surface texture. Here, we investigate the path type and path interval through simulations and experiments.

Material removal characteristics of densely spaced path lines are analysed here. Since the width of the polishing trace of each line is wider than the gap (pitch) between adjacent path lines, the removal at each location of a surface is the cumulated effect from all path locations within one-half of that width. For simplicity, the processing condition in all the lines are assumed constant and the path lines are evenly spaced straight lines on a planar surface. The case is similar to the removal in the inner part of a surface processed by scanning (Fig. 3.38a).

As a tool follows a tool path, the material removal profile with respect to a path point shows how the depth of removal varies along a direction orthogonal to the tool path tangent. If the tool orientation is constant and the tool follows straight path lines on a planar surface, the removal profile is constant along each and all the path lines.

The material removal map is constant along the path-line direction but undulates in the orthogonal direction. The undulation is a consequence of the overlapping of the removal profiles associated with adjacent path lines.

For the case of constant processing conditions and tool moving along a straight line, under the assumption of ellipsoidal contact, the shape of the removal profile is parabolic [9, 10]. The width of the profile is the width of the projection of the contact ellipse onto a line orthogonal to the direction of motion, and the centre of the profile is in the middle of the projection. The profiles corresponding to the tool moving along *X* and *Y* directions are shown in Fig. 3.39.

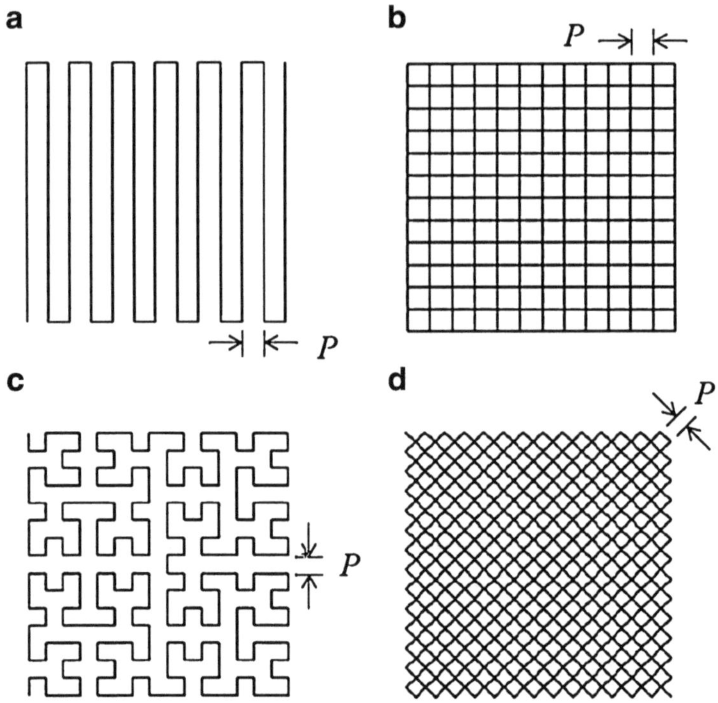

Fig. 3.38 Polishing tool paths: (**a**) scanning path; (**b**) bi-directional scanning path; (**c**) Hilbert path; (**d**) Peano path

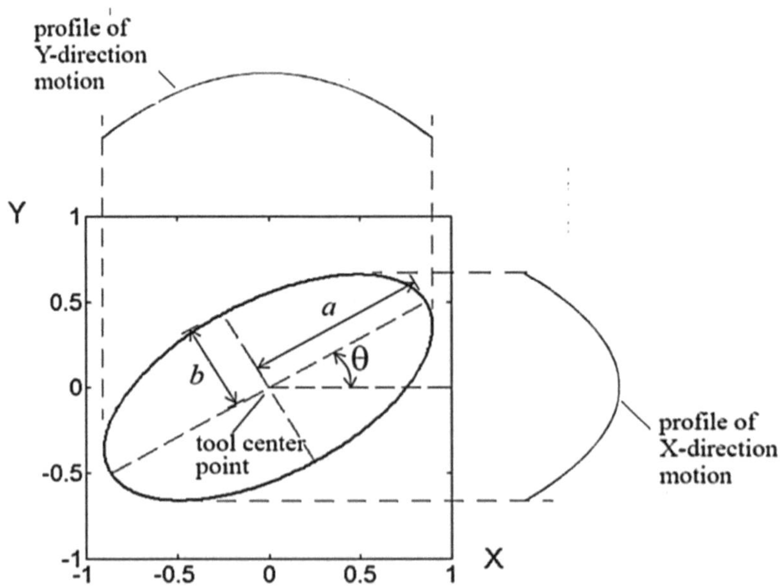

Fig. 3.39 Tool contact and removal profiles for tool motion along the *X* and *Y* directions

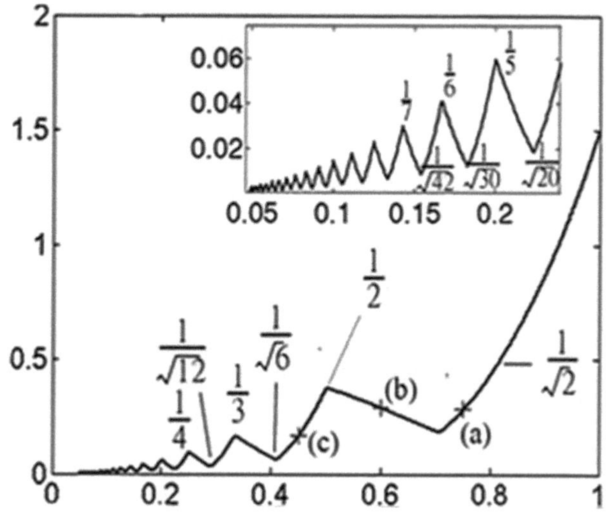

Fig. 3.40 Peak-to-valley height as a function of the pitch

Formulating the removal map may be simplified as a one-dimensional problem which can be solved numerically for given pitch values. Suppose the removal map is normalized such that the average depth of removal is unity. The height difference between the highest and lowest point of the normalized map is the normalized peak-to-valley (PV), h_{pv}. This parameter constitutes a way to assess the uniformity of the removal process. The h_{pv} values for normalized pitch p-values between 0.05 and 1 have been calculated and are plotted (Fig. 3.40). The pitch has been normalized with respect to the width of the removal profile. The major trend is that h_{pv} decreases with the pitch. However, locally, a smaller pitch can actually lead to a higher h_{pv}. The local maxima and minima in the plot are attributed to the complicated overlapping of adjacent removal profiles. The h_{pv} value initially decreases from 1.5 as p decreases from 1 where the path pitch equals the width of the removal profile. A cross-section of the removal map orthogonal to the path direction contains primary and secondary peaks (Fig. 3.41). For $1/\sqrt{2} < p < 1$, the primary peaks are the top of the removal profile of the individual path lines, and a secondary peak emerges midway between two adjacent path lines because of the overlapping of the adjacent profiles (Fig. 3.41a). The height of the secondary peaks increases relative to those of the primary peaks as p decreases. The first minimum of the h_{pv} plot appears when the height of the secondary peaks is equal to that of the primary peaks at $p = 1/\sqrt{2}$. For $1/2 < p < 1/\sqrt{2}$, the secondary peaks are higher than the primary peaks (Fig. 3.41b), and the primary peaks are not visible when $p = 1/2$. As p is further reduced, adjacent profiles also contribute to the height of the primary peaks and they become visible again (Fig. 3.41c). The locations of the three cross-sections in Fig. 3.41 are also labelled in the h_{pv} plot (Fig. 3.40).

The calculations suggest that the local maxima of the h_{pv} plot locate at $p = 1/n$, where n assumes integral values, and the local minimum between the adjacent maxima at $1/n$ and $1/(n + 1)$ is at $p = 1/\sqrt{n(n+1)}$. It seems that the primary and

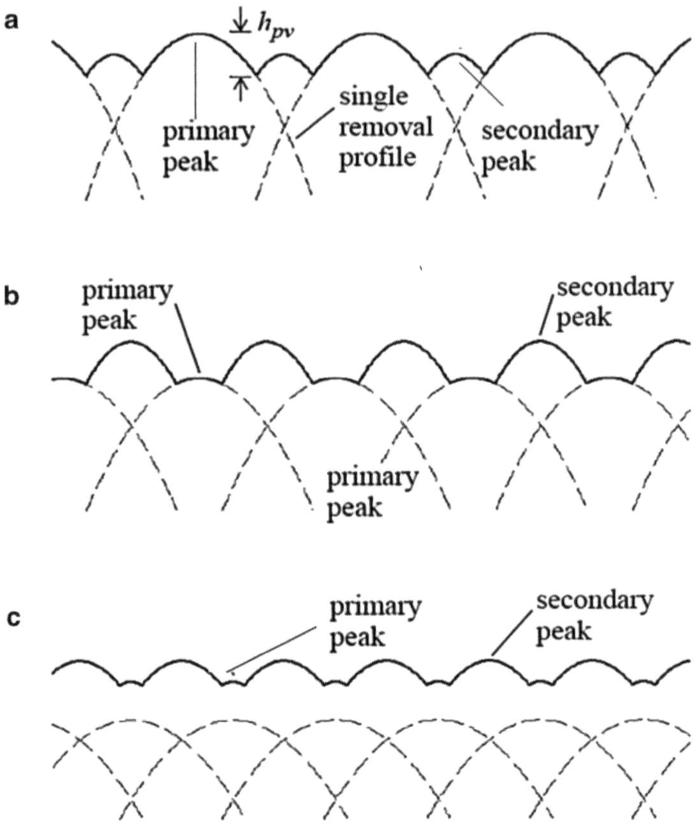

Fig. 3.41 Cross-section of the removal map. (**a**) $1/\sqrt{2} < p < 1$; (**b**) $1/2 < p < 1/\sqrt{2}$; (**c**) p slightly below 1/2

secondary peaks have the same height at the minima of the plot. Their relative heights change with p. The primary and secondary peaks take turns to be the higher peaks in successive ranges of pitch values between the minima.

Parallel lines of crests and troughs appear on the removal map.

Within a range of p over which the secondary peaks of the removal map are higher, the crests of the removal map are located midway between the path lines and the troughs at the path lines. As the pitch is reduced, the secondary/primary peaks grow in height with respect to the primary/secondary peaks where the slope of the plot is negative/positive. The secondary peaks disappear at the maxima of the h_{pv} plot where $p = 1/n$ and n is even.

Within a range over which the primary peaks are higher, the crests are located at the path lines and the troughs midway between the path lines. As the pitch is reduced, the primary/secondary peaks grow in height with respect to the secondary/primary peaks where the slope of the plot is negative/positive. The secondary peaks disappear at the maxima of the h_{pv} plot where $p = 1/n$ and n is odd.

Table 3.9 Removal comparison—circular contact

		Path density = 1/pitch (normalized)					
Tool path		2	4	6	8	12	Average
Scanning	h_{pv}	0.37	0.094	0.042	0.024	0.010	
	h_e	0.62	0.15	0.068	0.037	0.016	
	h_e/h_{pv}	1.68	1.59	1.62	1.54	1.60	1.6
Bi-scanning	h_{pv}	0.37	0.094	0.042	0.024	0.010	
	h_e	0.43	0.11	0.052	0.029	0.013	
	h_e/h_{pv}	1.16	1.17	1.24	1.21	1.30	1.2
	h_c	0.47	0.13	0.061	0.035	0.015	
	h_c/h_{pv}	1.27	1.38	1.45	1.46	1.60	1.4
Hilbert	h_{pv}	0.67	0.23	0.095	0.063	0.027	
Peano	h_{pv}	0.38	0.095	0.041	0.023	0.010	

How the polishing path pattern affects the uniformity of material removal is investigated in this section. Four tool path patterns are covered (Fig. 3.38). The removals in the inner surface as well as near the edges are examined.

Formulating the removal map arising from the convolution of the tool path and the tool impression are treated as two-dimensional problems. The tool impression function and the removal map are expressed as meshes of the same mesh size. The tool path is represented as a list of path points. The step size specifies the distance between the adjacent path points. Tool motion along the path is approximated by tool dwell at the path points. For a constant tool feed, the tool dwell times at all the path points are set equal. The removal map is computed by summing up the removal due to tool dwell at all the path points. The resolution of the simulations is affected by both the mesh size and the step size. They are set no greater than one-tenth of the path pitch, which itself is no greater than the width of the removal profile.

3.4.2 *Circular Tool Impression*

Consider the case of ellipsoidal tool impression in which the contact between and tool and the surface is circular. Under the assumptions stated in the section on material removal, this is approximately true if the tool is spherical and the radius of the tool is small compared with the radii of curvature of the surface. Without loss of generality, the diameter of the circular contact, which is equal to the width of the removal profile, is normalized to 1.

3.4.2.1 In the Inner Surface

For the four path patterns in Fig. 3.38, material removal maps were computed for path density ranging from 2 to 12, and the results are tabulated in Table 3.9 and plotted in Fig. 3.42. Path density (normalized) is the inverse of the normalized path

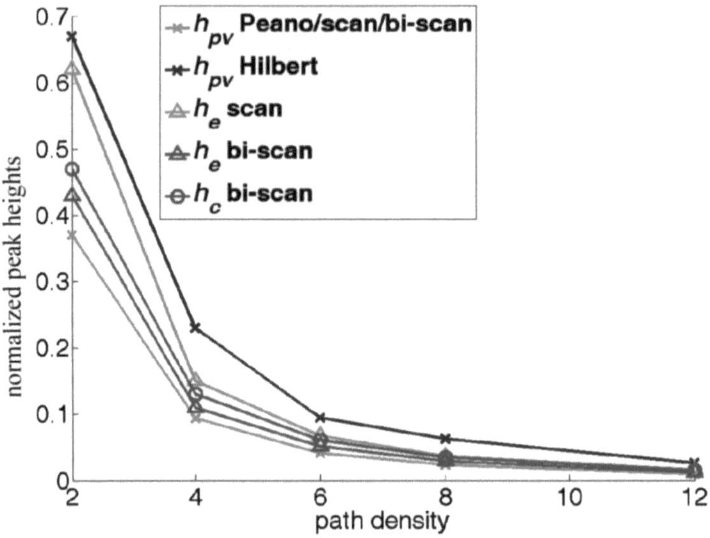

Fig. 3.42 Normalized peak heights vs. path density, circular tool

pitch *p*. The average height of the inner surface has been normalized to unity. PV height (h_{pv}) refers to the height of the highest peak to the lowest valley in the inner surface. (See the Sect. 3.4.2.2 below for definitions of edge and corner peak heights, h_e and h_c.)

From Table 3.9, h_{pv} decreases as the path density is increased for all the path patterns. This is consistent with the analysis of the overlapping effects in the last section—increase in the path density leads to more overlapping of adjacent removal profiles and uniformity in material removal.

The h_{pv} values for scan, bi-scan and Peano paths are practically the same. The texture of the removal maps are, however, quite different for these surfaces, as illustrated in the removal maps (Fig. 3.43) computed for the path density of 6.

Scanned surfaces contain evenly spaced troughs alternated with crests (Fig. 3.43a). The troughs lie on the scan lines and the crests between the scan lines because of the overlapping of adjacent removal profiles, and in agreement with the last section. The h_{pv} of the removal maps are also consistent with the values in the last section (Fig. 3.40). These help to confirm the results in both sections.

Bi-scanned surfaces contain isolated peaks and valleys in rectangular grids and the grid lines align with the two path-line directions (Fig. 3.43b). The removal map from bi-scanning may be seen as the result of scanning along two orthogonal directions. The amount of material removed in each of the component scans is the same. The height difference between the crests and troughs due to scanning in each direction is exactly one-half of the h_{pv} value from (uni-)scanning. Thus, the h_{pv} value from bi-scanning and (uni-)scanning are the same, and the peaks of bi-scanning are at the intersections of the crests of the two scan patterns and the valleys at the intersections of the troughs.

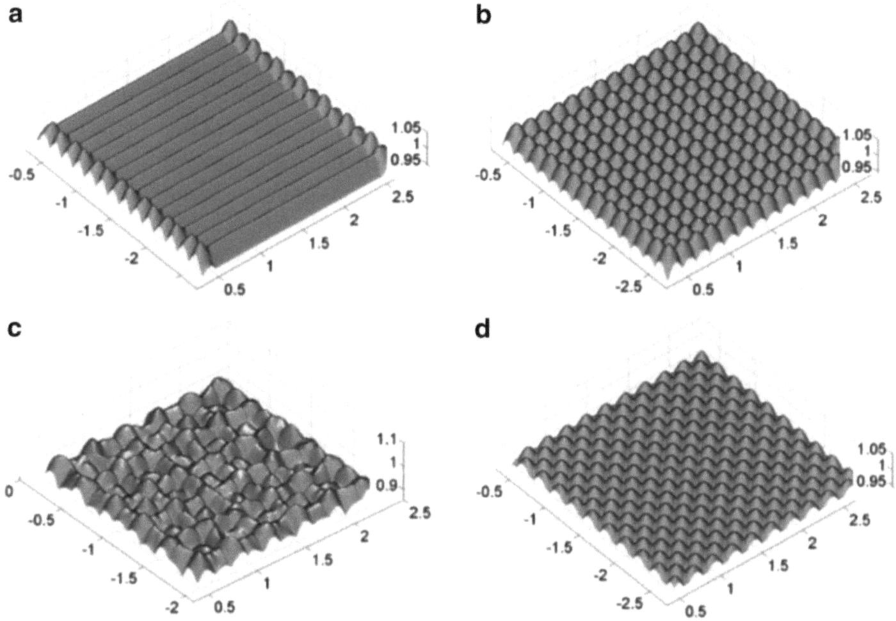

Fig. 3.43 Removal map for path density of 6: (**a**) scanning; (**b**) bi-scanning; (**c**) Hilbert; (**d**) Peano

Away from the edges, the grids produced by the Peano pattern and bi-scanning look the same, although they are formed quite differently. The size of the grids is the same but the orientations are 45° apart. Hence, the removal map from the Peano pattern also contains rectangular grids of peaks and valleys, and the h_{pv} value from the Peano pattern and that from bi-scanning is more or less the same (Fig. 3.43d).

The Hilbert path has the highest h_{pv} among all. On average, h_{pv} for the Hilbert path is more than double that of the other three path patterns (Fig. 3.42). Figure 3.43c shows a removal map from the Hilbert path. One can see irregular ridges along the path-line directions. Regions surrounded by ridges are slightly elevated in the middle. High points of the map are isolated peaks at the intersections of ridges. Low points are located near the bottom of the ridges. Although both the Hilbert and Peano paths are surface filling, self-similar, and the path segments are well balanced in the two orthogonal path directions, in terms of PV distance, the Hilbert path is actually inferior.

3.4.2.2 Near the Edges

Besides the texture of the interior region of the removal map, one observes high peaks at the edges of some of the maps. Edge peak height (h_e) and corner peak height (h_c) refer to the height of the highest peak along the edges (corners excluded) and around the corners from the lowest valley in the inner surface, respectively.

A line of peaks of alternating heights is present at the edges of the removal map along the scan direction (Fig. 3.43a). The peaks are located at the two ends of the crest lines. They can be attributed to the short path segments near the edges of the scan pattern which form the connections of adjacent scan lines. The height alterations of the peaks are due to the alternate presence and absence of those path segments along the edges. The edge contrast, defined by the h_e/h_{pv} ratio, is an indicator of how significant those edge peaks are compared to undulation in the inner surface. The edge contrast, also tabulated in Table 3.9, is about 1.6 on average for scan paths. If the total polishing time remains unchanged, the amount of time the tool stays at the edge segments is reduced as the path density is increased, implying the edge effects should also be reduced. However, h_{pv} also decreases as the path density is increased. Thus, while both the edge peaks and the undulation in the inner surface shrink as the path density goes up, the edge contrast seems to remain relatively significant.

The edge effect is also present in removal maps polished through bi-scanning. Figure 3.43b shows that high edge peaks are present at all the edges, as edge segments are present in all the edges of bi-scan patterns. The edge contrast for various path densities is tabulated in Table 3.9. The average edge contrast is about 1.2, which is smaller than that in (uni-)scanning. This can be attributed to (1) each of the component scans in bi-scanning has the removal strength of half of the (uni-)scan if the overall removal is the same in both; and (2) the edge effect at each edge of the removal map of bi-scanning comes from one of the component scans only.

The corner peaks stand out from the edge peaks (Fig. 3.43b) in bi-scanning. The corner contrast, as defined by the h_c/h_{pv} ratio (Table 3.9), serves as an indication of the relative height of the corner peaks. The average value of the corner contrast is about 1.4. On average, the height difference between a corner peak and an inner peak is roughly double that between an edge peak and an inner peak. A likely explanation is that the edge effects at both adjacent edges are present at a corner.

No edge effect is observed from removal maps polished using the Hilbert paths (Fig. 3.43c), although there are many short path segments along the edges of the path pattern. Notwithstanding, peaks of irregular heights show up near the edges as well as dispersed in the inner part of the removal maps. Some of the highest peaks are close to the edges as some are in the inner part. The same phenomenon is also observed when polishing using the Hilbert paths of other path densities. The formation of peaks is due to the local path pattern, which, at locations near the edges, is similar to that in the inner part for Hilbert patterns. Thus, it is reasonable that high peaks appear close to the edges as well as in the inner part.

Both Hilbert and Peano paths are surface-filling paths. As in Hilbert paths, edge effect is also not observed in removal maps polished using Peano paths (Fig. 3.43d). Peaks near the edges are not higher than the inner peaks. Both Peano and bi-scanning have similar path patterns in the interior. While the short path segments lining up along the edges contribute to the high edge peaks in bi-scanning, such path segments are not present in Peano patterns.

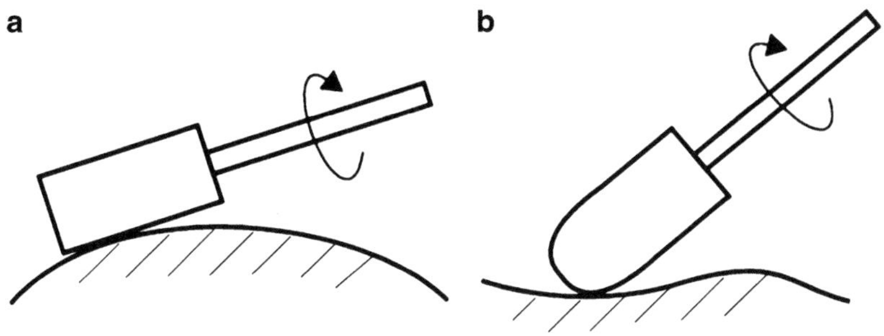

Fig. 3.44 Non-spheric polishing tools: (**a**) cylindrical; (**b**) filleted

3.4.3 Elliptic Tool Impression

Non-spheric tools may be used for polishing, depending on the application and the geometric form of the surface (Fig. 3.44). In general, the tool impression function is approximately ellipsoidal and the contact region elliptic under the assumptions in the section of material removal, if the tool is non-spherical. The width and depth of the removal profile varies depending on the orientation of the contact ellipse as well as on the direction of the tool feed.

Simulation results of removal are reported in Table 3.10 for polishing with the four tool path patterns. They cover tool orientations of $\theta = 0°$, 45° and 90°. Here, the major axis of the ellipse is aligned with the X-axis of the specimen reference frame when the orientation is 0° for a contact. A nominal contact ellipse with ellipticity (the ratio of the major and minor radii) of 2 was adopted. The size of the contact was also normalized so that the major diameter was 1.

The removal tends to be more uniform if the path density is increased (Table 3.10), as in polishing with the circular tool. For the same path density, the uniformity of removal for the elliptic tool is worse than that for the circular tool, and the tool orientation also matters. One expects the uniformity of removal for the elliptic tool to further worsen if ellipticity is further increased.

3.4.3.1 Scanning Path

Distinct crests and troughs can be seen on the removal map from scanning with elliptic tools as with circular tools. When the tool orientation is 0°, the major direction of the contact ellipse is orthogonal to the scan lines, and the h_{pv} values are practically the same as those of the circular tool. They do not seem to change significantly as the tool orientation is changed from 0° to 45°. The 90° tool has the highest h_{pv} values and, on average, they are about 3.9 times those of the 0° tool. Although, 0° and 45° tools have similar h_{pv} values, the texture of the removal maps are quite different. For instance, the crests of the map of the 0° tool are wider and

Table 3.10 Removal comparison—elliptic contact

Tool path			Path density = 1/pitch (normalized)			Average
Type	Orientation		4	6	8	
Scanning	0°	h_{pv}	0.0985	0.0418	0.0236	
		h_e	0.28	0.12	0.071	
		h_e/h_{pv}	2.83	2.86	2.96	2.9
	45°	h_{pv}	0.100	0.0375	0.0156	
		h_e	0.20	0.073	0.034	
		h_e/h_{pv}	2.00	1.92	2.13	2.0
	90°	h_{pv}	0.375	0.167	0.0938	
		h_e	0.48	0.19	0.11	
		h_e/h_{pv}	1.30	1.11	1.17	1.2
Bi-scanning	0°, 90°	h_{pv}	0.240	0.104	0.0587	
		h_e (high)	0.30	0.13	0.079	
		(low)	0.27	0.11	0.064	
		h_e/h_{pv}	1.25	1.30	1.34	1.3
		h_c	0.33	0.14	0.084	
		h_c/h_{pv}	1.38	1.40	1.42	1.4
	45°	h_{pv}	0.100	0.0375	0.0156	
		h_e	0.12	0.052	0.023	
		h_e/h_{pv}	1.20	1.37	1.44	1.3
		h_c	0.15	0.067	0.030	
		h_c/h_{pv}	1.50	1.76	1.88	1.7
Hilbert	0°, 90°	h_{pv}	0.55	0.25	0.15	
	45°	h_{pv}	0.30	0.17	0.10	
Peano	0°, 90°	h_{pv}	0.102	0.0384	0.0157	
	45°	h_{pv}	0.236	0.104	0.0586	

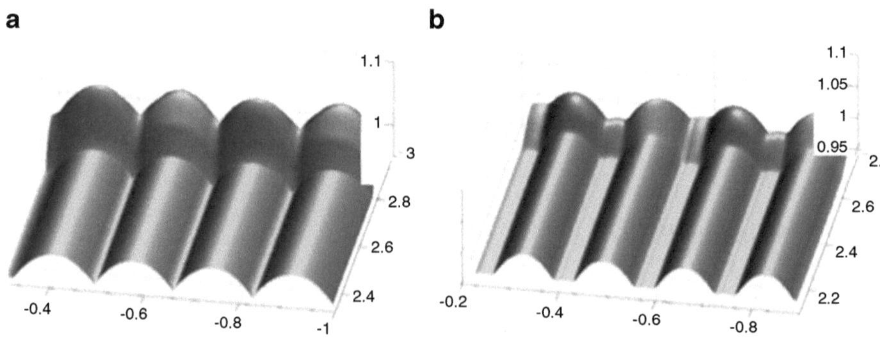

Fig. 3.45 Removal map from scanning using elliptic tool at path density of 6: (**a**) $\theta = 0°$; (**b**) $\theta = 45°$

the troughs are narrower compared with those of the 45° tool (Fig. 3.45). To facilitate comparisons, the PV changes are also plotted in Fig. 3.46 using values from Tables 3.9 and 3.10.

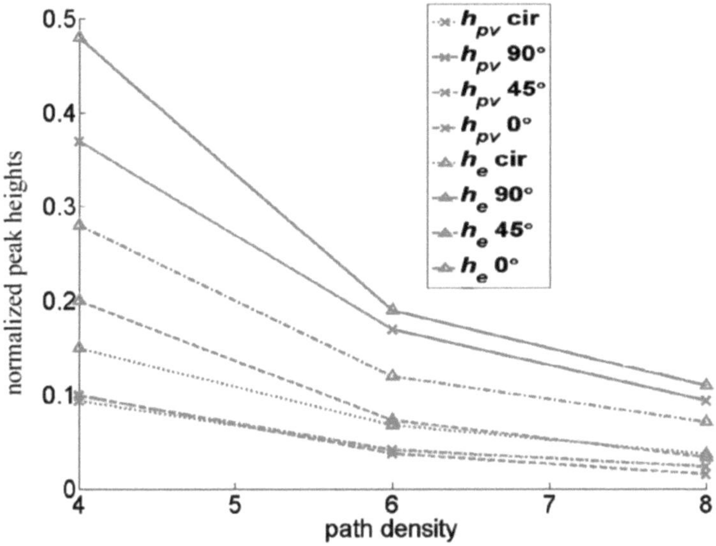

Fig. 3.46 Normalized peak heights vs. path density for scanning path

The h_{pv} value depends on the ratio of the pitch to the width of the tool removal profile. The 0° tool and the circular tool have the same h_{pv} values as they have the same profile width along the scan lines. The h_{pv} values are approximately the same for the 90° tool at a path density of 8, and for the 0° tool at a path density of $\tilde{4}$. In these two cases, the ratio of the pitch to profile width are the same, as both the path pitch and the profile width of the former are half of those values of the latter.

The h_{pv} curve in Fig. 3.40 provides accurate indications of the h_{pv} values for scanning with elliptic tools as with circular tools, with proper adjustment of the pitch value based on the profile width of the elliptic tool. To obtain the correct h_{pv} value for elliptic tools from the figure, the original pitch of the scan path should be divided by the profile width if it is not unity. For instance, the profile width is $1, 0.79$, and 0.5 for tool orientations $0°, 45°$ and $90°$, respectively. The h_{pv} values for the path density of 4 (path pitch of 0.25) should be based on pitch values of 0.25, 0.32, and 0.5. The h_{pv} values thus obtained from Fig. 3.40 are in fact rather close to the h_{pv} values in Table 3.10.

As the tool orientation is increased from 0° to 90°, the removal profile associated with the individual scan lines becomes narrower and deeper. One expects the h_{pv} values to go up when this happens, and this in fact is consistent with the simulation results of the 0° and 90° tools. The removal profiles become slightly narrower and deeper when the tool orientation is changed from 0° to 45°. Yet, corresponding changes in h_{pv} values among the various path densities are rather haphazard (Table 3.10). This is probably due to the complicated overlapping of adjacent removal profiles and the sign changes of the slope of the h_{pv} curve in Fig. 3.40.

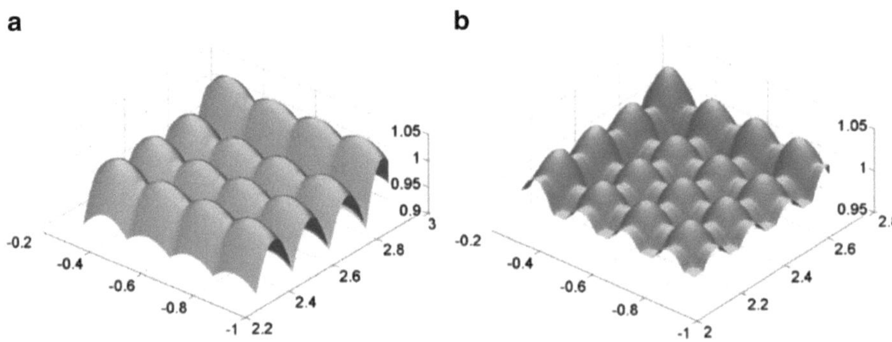

Fig. 3.47 Removal map from bi-scanning using elliptic tool at path density of 6: (**a**) $\theta = 0°/90°$; (**b**) $\theta = 45°$

For all the tool angles, a line of isolated peaks can be seen along each of the two edges, with short path segments lining up those edges. Edge effect is more evident in the use of elliptic tools compared to using circular tools. The h_e values for the elliptic tools are, in general, higher than the corresponding h_e values for the circular tool for the same path density (Fig. 3.46). The 90° tool has the highest edge peaks.

As to the height difference ($h_e - h_{pv}$) between the edge peak and the inner peak attributable to the connecting segments of scan paths, it seems $h_e - h_{pv}$ is most significant for the 0° tool, then the 45° tool, and the 90° tool in turn (Fig. 3.46). This is likely because of the lining up of the major/minor axes of the contact ellipse with those connecting segments when the contact orientation is small/large.

The edge contrast is relatively large when the contact orientation is close to 0°, and small when the orientation is close to 90°. The average edge contrast is 2.9, 2.0, and 1.2 for the 0°, 45° and 90° tools, respectively. The 0° tool has the highest edge contrast as the size of $h_e - h_{pv}$ is large. The 90° tool has notably low edge contrast values, in part because the $h_e - h_{pv}$ values are relatively small, but more importantly because the h_{pv} values of the 90° tool are relatively large.

3.4.3.2 Bi-scanning Path

Bi-scanning consists of scanning in two orthogonal directions. For scanning with an elliptic tool, the removal profiles along the two scan directions may have different width and height, depending on the tool orientation.

The 0° and 90° tools produce patterns on the removal map which differ only by a 90° rotation. The crests and troughs along one of the scan directions are more distinct compared with those along the other direction because of the disparities in the removal profiles along the two directions, and giving rise to the directional texture of the removal map (Fig. 3.47a). The crests and troughs along both scan directions have the same form and size for the 45° tool, resulting in rectangular grids of peaks and valleys in the removal map (Fig. 3.47b). The h_{pv} values of the

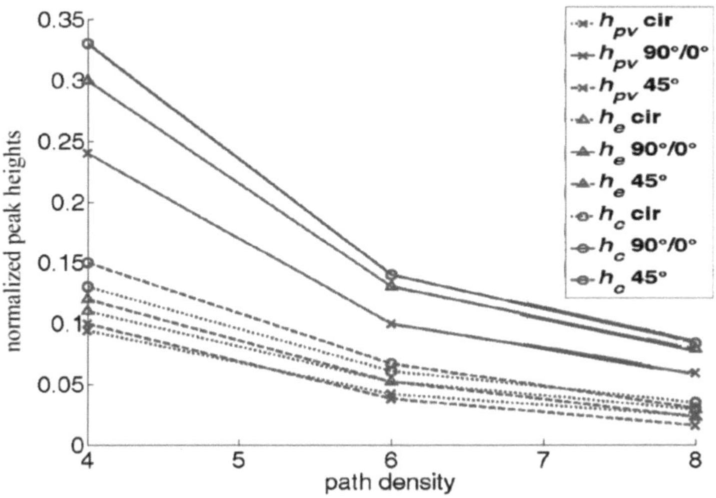

Fig. 3.48 Normalized peak heights vs. path density for bi-scanning path

0°/90° tool are about 2.5 times those values of the circular tool; those of the 45° tool are lower and are slightly less than those values of the circular tool (Tables 3.9 and 3.10; Fig. 3.48).

The two scan components of bi-scanning with a 0°/90° tool are equivalent to (uni-)scan with a 0° tool in one direction and a 90° tool in the orthogonal direction, both in half strength. Thus, the h_{pv} values of bi-scanning with a 0°/90° tool are between those of (uni-)scanning with 0° and 90° tools. On average, they are about 2.4 times those of (uni-)scanning with 0° tools and 0.6 times those of (uni-)scanning with 90° tools. As for bi-scanning with a 45° tool, both component scans are comparable to (uni-)scanning with a 45° tool. The h_{pv} values of bi-scanning and (uni-)scanning with the 45° tool are the same.

In general, one expects the removal maps to differ by a 90° rotation for tool orientations $45° \pm \varphi$, where $0° \leq \varphi \leq 45°$. The surface texture is directional. The directionality diminishes as φ comes near 0°. The h_{pv} value of bi-scanning with a $45° \pm \varphi$ tool is in between the h_{pv} values of (uni-)scanning with $45° + \varphi$ and $45° - \varphi$ tools. In particular, the h_{pv} values for bi-scanning and (uni-)scanning with the 45° tool are the same.

Edge effects are also evident in bi-scanning with elliptic tools, as indicated by the presence of higher peaks along the four edges of the removal maps. Edge peaks with elliptic tools are higher than the corresponding edge peaks with the circular tool. Comparisons of edge peaks are shown in Tables 3.9, 3.10, and Fig. 3.48. Edge peaks of the 0°/90° tool are the highest. While edge peaks of the removal map have the same height when polished with the 45° tool, edge peaks along one pair of opposing edges are higher than those along the other pair when polished with the 0°/90° tool (Table 3.10) because of the disparity in the removal profiles along the two path directions in the latter case.

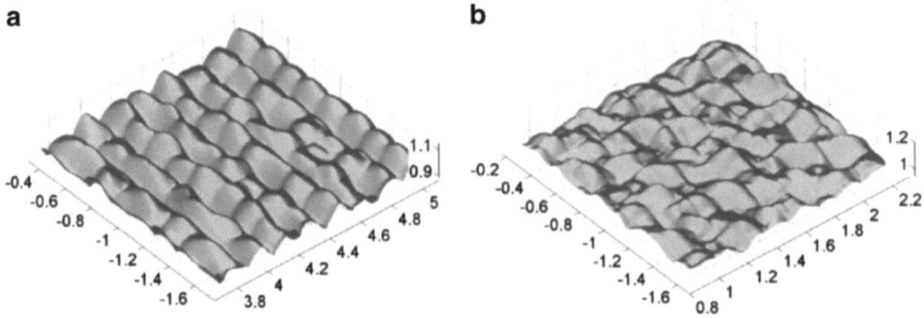

Fig. 3.49 Removal map from Hilbert using elliptic tool at path density of 6: (**a**) $\theta = 0°/90°$; (**b**) $\theta = 45°$

Height difference between the edge peaks and the inner peaks is significantly lower in bi-scanning than in (uni-)scanning. This is evident when the comparisons are made between the $h_e - h_{pv}$ values for the higher edges of the 0°/90° tool, the edges of the 45° tool, and the lower edges of the 0°/90° tool in bi-scanning, and the corresponding $h_e - h_{pv}$ values of the 0° tool, 45° tool, and 90° tool in (uni-) scanning.

The edge contrast seems to be lower in bi-scanning than in (uni-)scanning. The average edge contrast of the three tool orientations is about 1.3, which is low compared with the average of about 2.0 in (uni-)scanning. It is noted that the high edge contrast in 0° (uni-)scanning is not present in 0°/90° tool bi-scanning, which contains a component scan of the former. Quite likely this is due to the averaging out of the high edge peaks when scanning is executed in two orthogonal directions.

Corner peaks are higher than edge peaks in bi-scanning with elliptic tools. Corner peaks with elliptic tools are higher than the corresponding corner peaks with the circular tool. Corner peaks of the 0°/90° tool are the highest.

Corner contrast (h_c/h_{pv}) is high in elliptic tool bi-scanning. The average corner contrast is 1.71 and 1.40 for the 45° tool and 0°/90° tool, respectively; the latter is comparable to that in circular tool bi-scanning (1.43). The height difference ($h_c - h_{pv}$) between a corner peak and an inner peak is very close to the sum of the height difference ($h_e - h_{pv}$) of the adjacent edges, indicating the accumulation of edge effects at the corners.

3.4.3.3 Hilbert and Peano Paths

High edge peaks are not observed in the removal maps when either the Hilbert or Peano paths is used for polishing.

For polishing with the Hilbert path, the texture is visibly directional for contact orientations significantly different from 45°. The contour is quite pronounced along the major direction of the contact ellipse for the 0°/90° tool (Fig. 3.49a). Isolated peaks are oblong and some join to form ridges. Orthogonal ridges present when

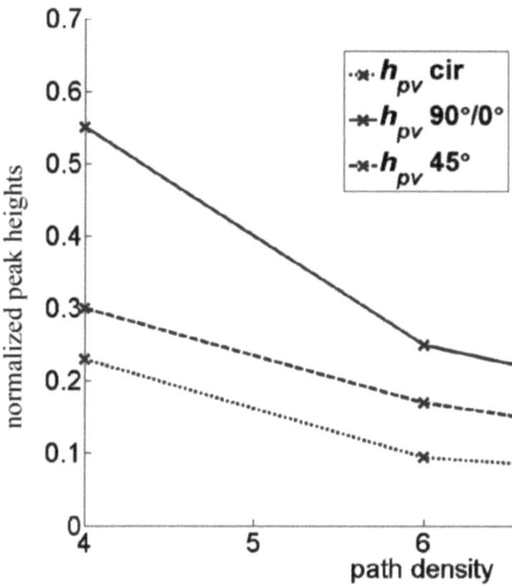

Fig. 3.50 Normalized peak heights vs. path density for the Hilbert path

polishing with circular tools become less impressive. On the other hand, elevated regions originally entangled within the ridges are more visible, and some are connected to also form minor ridges along the major direction of the contact ellipse. For the 45° tool, the ridges along the two path segment directions are still visible as in the case of circular tools (Fig. 3.49b). Some of the isolated peaks at the junction of those ridges are slightly elongated along the major direction of the contact ellipse.

The orientation of the contact affects the width of the removal profile and, thus, the h_{pv} values. The 0°/90° tool gives rise to higher h_{pv} values than the 45° tool, which are in turn higher than those of the circular tool (Fig. 3.50). A likely explanation is that the width of the removal profiles are largest for the circular tool, those of the 45° tool are less, and those along one of the two path directions are the least for the 0°/90° tool.

For polishing with the Peano paths, diagonal ridges and a grid of peaks are seen on the inner surface when using the 0°/90° tool (Fig. 3.51a), and elongated peaks connected to form ridges along diagonal lines (along the major direction of the contact ellipse) are seen when using the 45° tool (Fig. 3.51b). PV values (h_{pv}) are plotted in Fig. 3.52. The 0°/90° tool and the circular tool have similar h_{pv} values. Unlike the case of Hilbert paths, the 45° tool has the highest h_{pv} values among all tool orientations for polishing with Peano paths. This, again, is because the major direction of the contact ellipse is aligned with half of the path segments for this tool orientation. All these observations of polishing with Peano paths is consistent with the fact that, in the inner part, the patterns formed by the Peano path and bi-scanning path differ only by a 45° rotation. For instance,

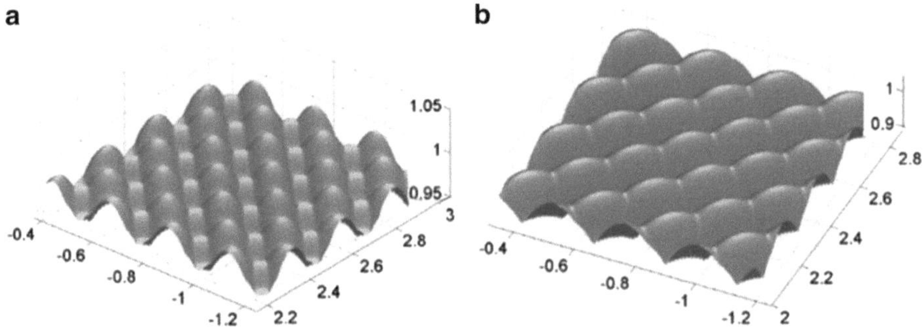

Fig. 3.51 Removal map from the Peano path using elliptic tool at path density of 6: (**a**) $\theta = 0°/90°$; (**b**) $\theta = 45°$

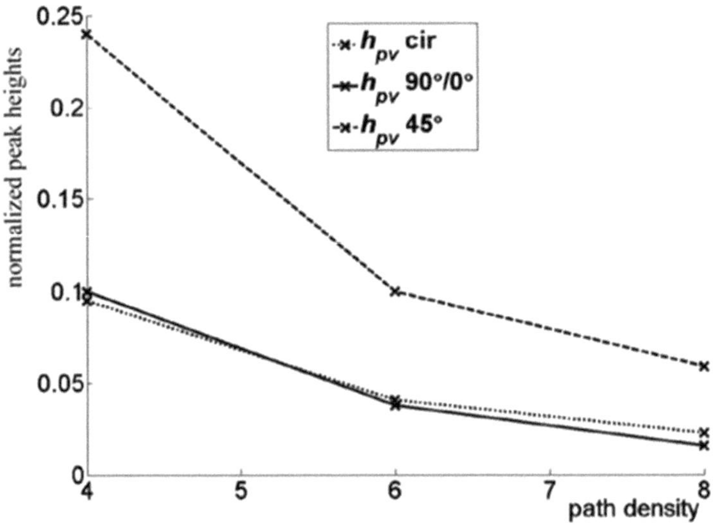

Fig. 3.52 Normalized peak heights vs. path density for Peano path

the texture and h_{pv} value from polishing with a $45° \pm \varphi$ tool following a Peano path are the same as those from polishing with a $\varphi/90° - \varphi$tool following a corresponding bi-scanning path.

The h_{pv} values from Peano paths are, in general, smaller to those from Hilbert paths due to the relatively high irregularity of the Hilbert pattern, although both are surface-filling paths. On average, h_{pv} values from the Peano path using the 0°/90° tool are about 0.22 those from the Hilbert path using the 45° tool, and h_{pv} values from the Peano path using the 45° tool are about 0.4 those from the Hilbert path using the 0°/90° tool (Table 3.10).

3.4.4 Experiments

Polishing experiments were conducted to investigate how the tool path affects the uniformity of removal. The focus was on the edge effects. Comparisons were made between simulations and experiments.

Specimens of aluminum plates were prepared and the surface flatness was within 2–3 μm. A relatively soft material was selected as the specimen material to promote the rate of removal and high signal-to-noise ratio of surface measurement. Filleted tool sticks were adopted for the polishing experiments. The tools contain alumina mixed in a binding material of synthetic rubber. The tools were selected for rate of removal demonstration rather than for surface roughness reduction.

During experiments, the specimen was fixed and relative motion between the tool and the specimen was obtained by a numerical control motion platform. The tool spindle was free to slide along the tool axis, which was fixed at a small inclination angle from the vertical axis. The tool orientation was kept constant during polishing and a nominal spindle speed was used. A constant polishing force was obtained by deadweight. Lubricant/thinner was supplied to the polishing zone to flush out polishing debris and to carry away the generated heat.

Both the Scanning path and Peano path were followed for polishing. The polished surfaces were measured using a Form Talysurf profiler. The measurement results for the two tool paths are shown in Figs. 3.53 and 3.54, respectively.

3.4.4.1 Scan Path Polishing

Tool axis was fixed such that its projection on the surface was along the east-west direction (i.e. orthogonal to the scan lines), which was also the principal direction of the tool contact region. The sliding direction of abrasive grits at the contact was along the north-south direction (i.e. same as the scan-line direction). The path pitch was set at 1 mm. The removal map obtained from surface measurement is shown in Fig. 3.53a. Relatively uniform removal is found in the inner surface. Connecting short path segments of the scan path are close to the north and south edges, where the tool moved from one scan line to another. Clear texture of crests and troughs, which are characteristic of scanning path polishing, can be seen in the removal map. Removal along the north and south edges are rather severe, indicating the severity of edge effects of scanning.

An east-west line profile is shown in Fig. 3.53b. The average depth of removal is approximately 15 m in the inner profile. The radius of the tool contact region along this direction is about 1.8 mm, as suggested by the ramp-up length. The jagged profile suggests inhomogeneity of the distribution of abrasive grit within the tool. Nonetheless, still clearly seen is the spatial frequency component at 1 mm wavelength, which corresponds to the pitch of scanning.

Figure 3.53c shows a line profile along the north-south direction. This profile is smoother than the line profile in Fig. 3.53b, possibly because this profile is along the

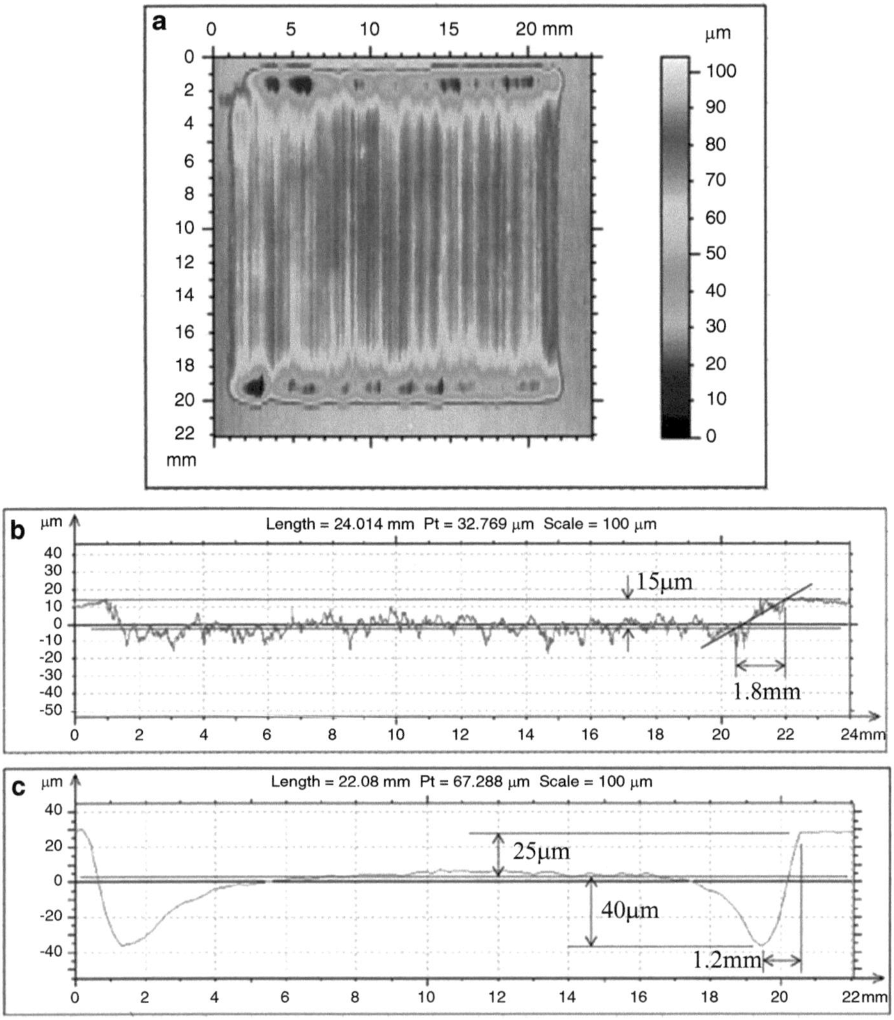

Fig. 3.53 Experimental results using the scanning path. (**a**) removal map; (**b**) line profile along east-west direction near the removal map middle; (**c**) line profile along north-south direction near the removal map middle

sliding direction of abrasive grit while that in Fig. 3.53b is along the cross-direction. The average depth of removal is approximately 25 m in the inner part, suggesting that the line profile is within a trough of the surface. The height of the edge peaks with reference to the inner region is about 40 m. The valleys near the two ends of the profile indicate the severity of removal and edge effects of scanning. The radius of the tool contact along this direction is about 1.2 mm, which is approximated by the distance from the bottom of one of the valleys to the edge.

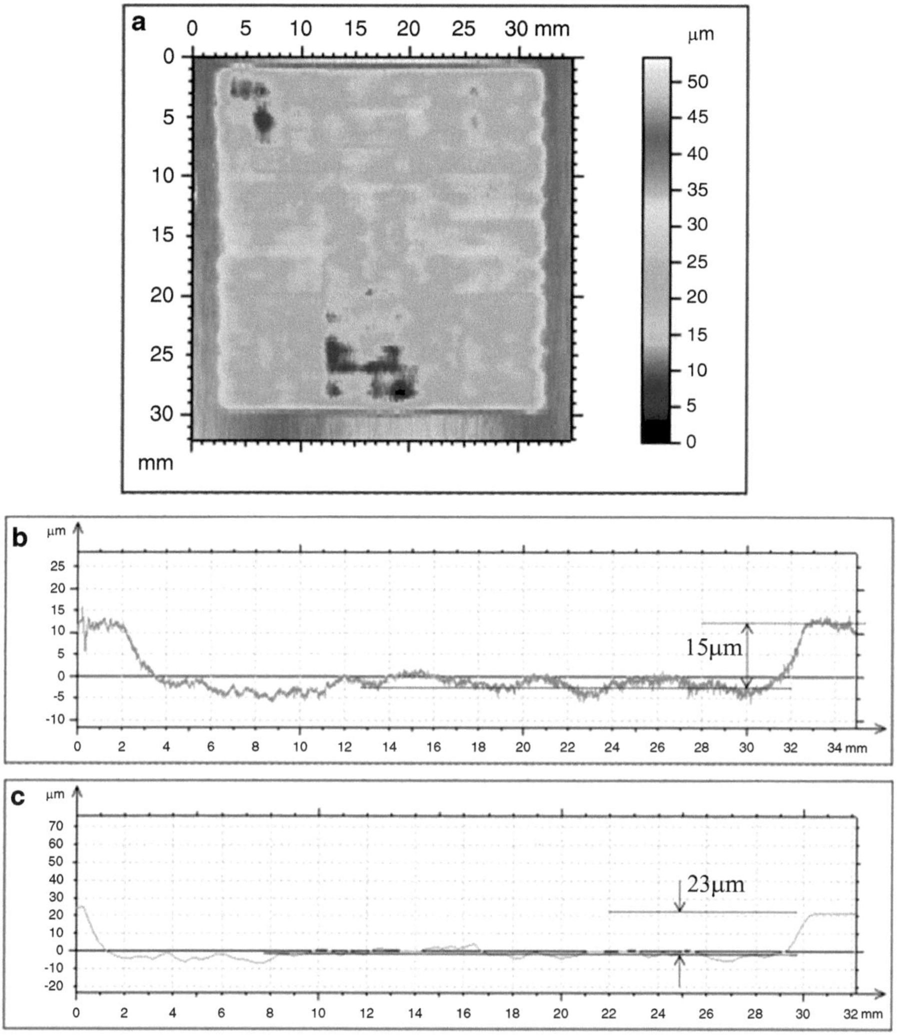

Fig. 3.54 Experimental results using the Peano path. (**a**) removal map; (**b**) line profile along east-west direction near the removal map middle; (**c**) line profile along north-south direction near the removal map middle

Severe removal of material near the two opposite edges is clearly seen in Fig. 3.53. The above line profiles suggest the ellipticity of the contact is about 1.8 mm/1.2 mm = 1.5, and the orientation of the contact $\theta = 0°$. The path density is approximately 2×1.8 mm/1 mm = 3.6. The reference case in Table 3.10 of ellipticity = 2, $\theta = 0°$ and path density = 4 suggests the height of the edge peaks to be in the order of 0.28×25 µm = 7 µm, which is small compared to the height of 40 µm in Fig. 3.53c. Imprecision in the measurement of the contact radii and the depth of removal, to some extent, contribute to experimental error. More probable sources of discrepancy are accumulation of debris and directionality of the removal rate. The tool motion changed abruptly near the edges, which could lead to accumulation of debris at the tool contact and affects the material removal near the edges. The movement of the tool relative to the surface at the contact is parallel to the scan lines and orthogonal to the short path segments at the edges. It is possible that the rate of removal was actually higher when the tool was on those short path segments which cause severe removal at the edges.

3.4.4.2 Peano Path Polishing

For polishing using the Peano path, tool orientation is the same, i.e. the projection of the tool axis on the surface is along the east-west direction, and the sliding direction of the abrasive grit along the north-south direction. The path pitch is 0.5 mm. The removal map obtained is shown in Fig. 3.54a. The removal of material seems to be rather uniform in the inner surface, except at a few isolated regions. Those isolated regions of extra removal are likely due to the inconsistent flushing of the debris, which was carried out manually during the experiment.

Line profiles along the east-west and north-south directions are shown in Fig. 3.54b, c. The first profile is relatively rough as it runs across the direction of grit sliding. The average depth of removal is about 15 m. The second is relatively smooth as it is along the grit sliding direction. The average removal is slightly above 20 m. Again, the depth of removal of the second profile could be sensitive to the location of the line as it is along the grit sliding direction.

Edge peaks seem to be not higher than the inner peaks according to the removal map (Fig. 3.54a) from Peano path polishing or from the line profiles (Fig. 3.54b, c). This is consistent with the simulation results in the last section. While the accumulation of debris and directionality of the removal rate are likely to add to the edge effect in scanning paths, they are unlikely to contribute significantly to the uniformity of removal for Peano path polishing. Unlike scanning paths in which the tool motion and tool movement direction change at the edges only, such changes are in fact distributed over the entire inner surface as well as near the edges for Peano paths.

3.5 The Influence of Tool Path on Mid-frequency Errors

3.5.1 Path Type to Mid-spatial Frequency Errors

To perform feasibility studies, the proposed tool path was used in two polishing platforms. The specimens were BK 7 in the experiments. The scope was limited to finding out whether the tool path was suitable for polishing.

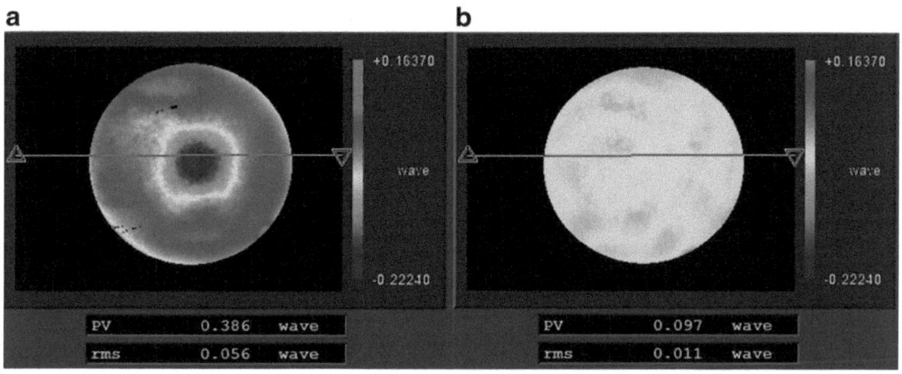

Fig. 3.55 Residual error map of circular specimen (**a**) before and (**b**) after MRF polishing based on the Peano-like path. MRF magnetorheological finishing

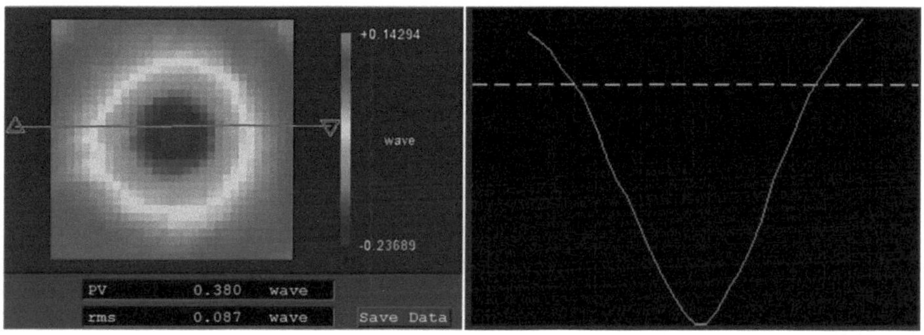

Fig. 3.56 The used removal function, with 0.38 λ/min peak removal rate

The first experiment was performed to validate the tool path method and to investigate the effects of a tool path on the polishing process. The specimen was a circular Φ100 mm parabolic specimen. The initial surface map was shown in Fig. 3.55 (upper). Polishing was based on a magnetorheological finishing (MRF) system made locally. The system adopted a planetary motion structure and the magnetic field was generated using permanent magnet. During experiment, the self-rotating speed and orbital speed were set at 300 revolutions per minute (rpm) and 60 rpm, respectively, which produced a Gaussian-like removal function (shown in Fig. 3.56) with a peak removal rate at 0.38 λ/min ($\lambda = 632.8$ nm); the function spanned a Φ12 mm region. The tool path (0.8 mm pitch, 3,120 points, shown in Fig. 3.57a) followed the proposed Peano-like path outside the vertex cap and square elements inside.

The resulting surface was measured as shown in Fig. 3.55 (lower) after a 145-min polishing run. Measurement revealed that the PV and root mean square (RMS) converged from 0.386λ and 0.056λ to 0.097λ and 0.011λ, respectively. The proposed path was characterized by a large number of sharp turns and short path

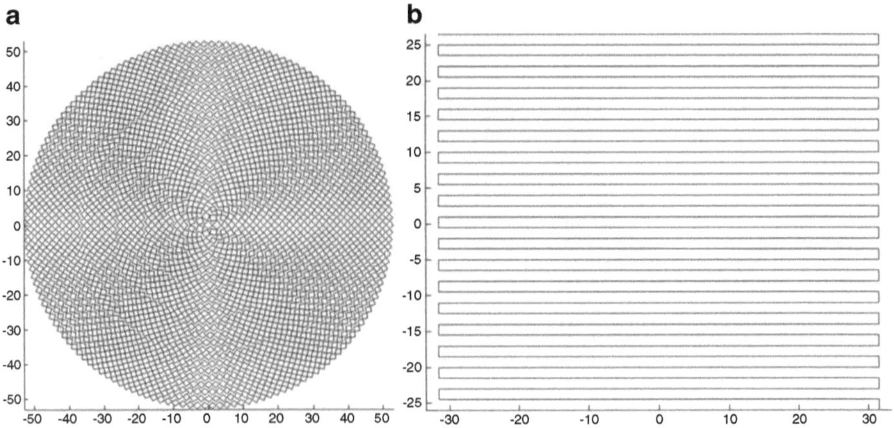

Fig. 3.57 The polishing path. (**a**) Peano-like path; and (**b**) the raster path

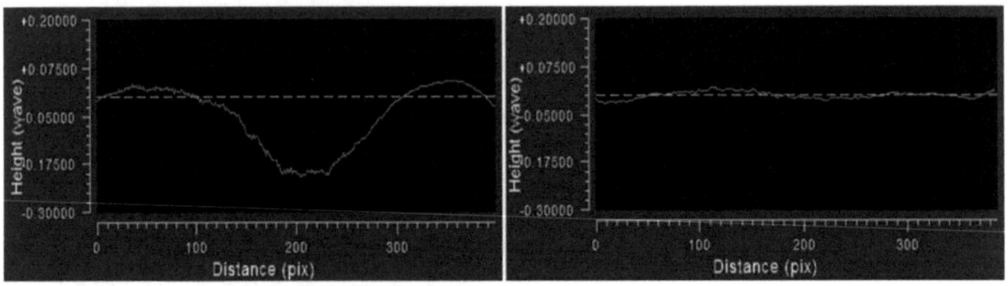

Fig. 3.58 In-section profiles of the error map before (*left*) and after (*right*) polishing

segments. The effective reduction of the PV and RMS provide confirmation that Peano-like paths are suitable for polishing using MRF, from the standpoint of machine dynamics and the polishing process. From the in-section profiles (Fig. 3.58) of the maps in Fig. 3.55, it could be seen that the dent near the center of the specimen was largely removed, which was consistent with the above statistical results.

Further assessment of the results was made based on the power spectral density (PSD) of the specimens before and after polishing (Fig. 3.59). Improvement of the surface was supported by the lowering of the PSD curve after polishing by more than one decade. The major peak in the vicinity of 0.04 mm^{-1} could be related to the form error of the surface. In the PSD curve (after), the minor peaks near 0.1 mm^{-1} and slightly above 1 mm^{-1} could be related to the removal impression of the tool and the path pitch respectively. The presence of this latter peak indicated that the tool path could cause mid-frequency ripples and the effects may not be negligible. The proposed tool path method supported uniform polishing, and traces of the polishing path were not obvious (Fig. 3.55 [lower]). If the tool path were not uniform, it was quite likely that removal at certain locations might be considerably

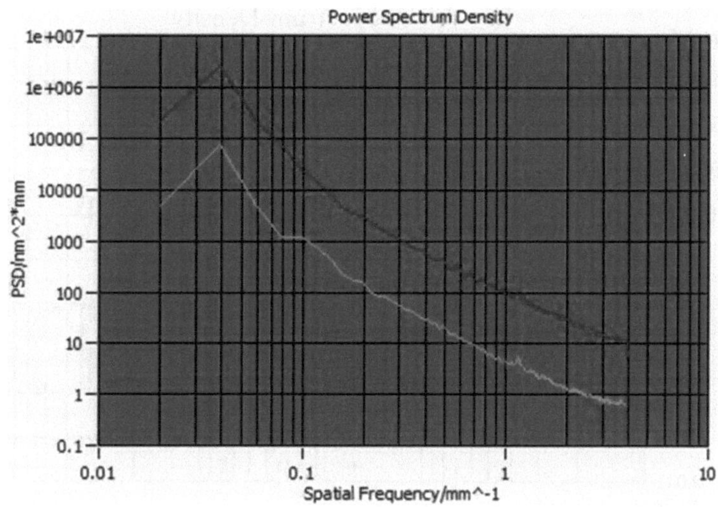

Fig. 3.59 PSD results before (*red curve*) and after (*green curve*) polishing. *PSD* power spectral density

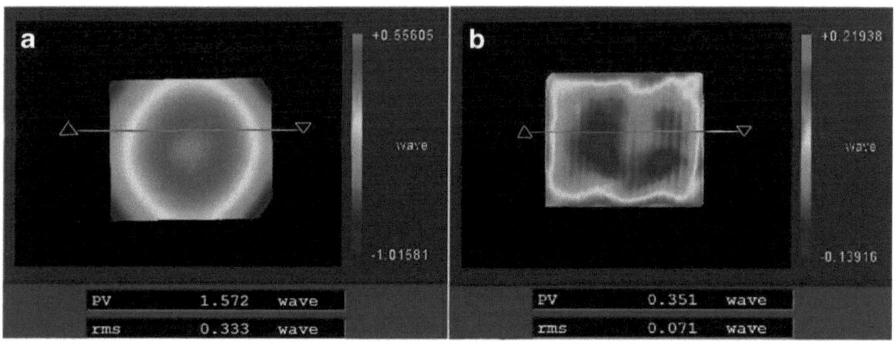

Fig. 3.60 Residual error map of rectangular specimen (**a**) before and (**b**) after MRF polishing based on scanning path. *MRF* magnetorheological finishing

higher than that at other locations, leading to bigger defects of those on the polished surface.

A follow-up experiment was conducted for tool path comparison. A rectangular parabolic specimen (50 mm $\times$ 60 mm) was polished using the scanning path shown in Fig. 3.57b. Polishing was carried out on the above MRF system. The tool removal function and path pitch were the same as before. Measurements of the surface were shown in Fig. 3.60. There was general reduction of the PV and RMS from 1.572λ and 0.333λ to 0.359λ and 0.072λ, respectively. Marks left by polishing based on scanning was evident (Fig. 3.60b). While the path pitch could have been

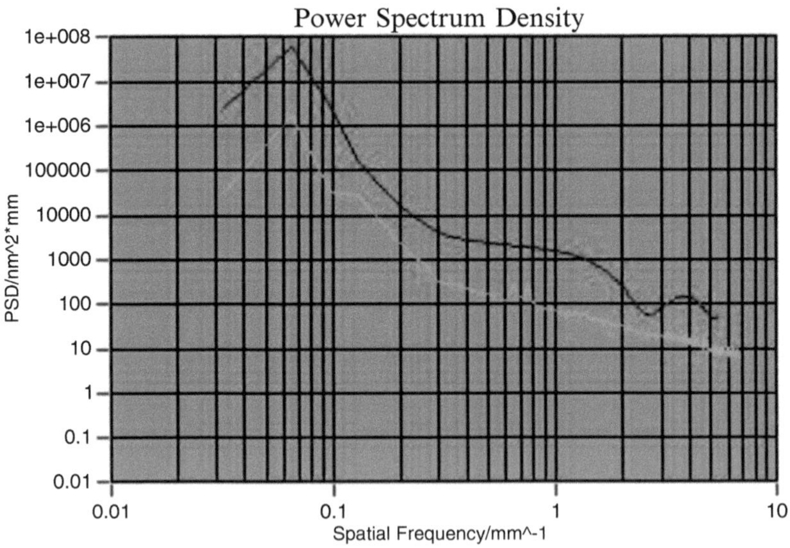

Fig. 3.61 PSD results before (*red curve*) and after (*green curve*) polishing for the second experiment. *PSD* power spectral density

reduced to suppress the generation of marks and ripples, the results indicated that, for the same path pitch, Peano-like paths could be more uniform than scanning paths. This may be related to the fact that, with Peano-like paths, the removal Gaussian went past each surface location along more distributed directions than with scanning paths. PSD curves of the rectangular specimen before and after polishing were also given (Fig. 3.61). Again, perturbations near 0.7 mm^{-1} could be related to the effects of the scanning pitch.

Another circular specimen was used to provide further validation. It was a parabolic surface with an aperture of 100 mm. Polishing was carried out using pitch and cerium oxide on a multi-axis polishing platform. The machine had an *X-Y-Z* working range of 600*620*260 mm and a 360° rotating turntable. It was equipped with a planetary tool motion to achieve Gaussion-like removal with a central high point. Another tilt axis enabled the tool axis to remain normal to the workpiece. The tool path followed the proposed tool path and the path pitch was set at 1 mm. The self-rotating and orbiting speeds of the polishing pad were 300 rpm and 60 rpm, respectively. The pressure on the pad was 0.08 MPa.

After 5.5 h of polishing, the shape of the surface was found to be not significantly affected by polishing. More importantly, it was confirmed by WYKO measurement that the surface Ra achieved was 0.91 nm (Fig. 3.62), which was rather good for optical lenses. This served as further confirmation that Peano-like paths were suitable for smoothing to reduce the surface roughness.

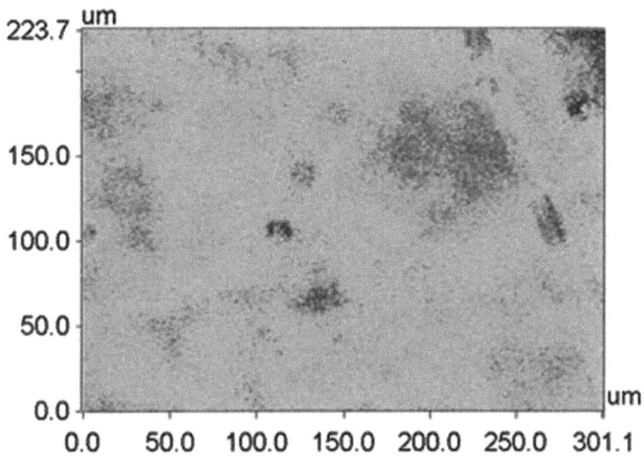

Fig. 3.62 Roughness measurement of circular specimen after pitch polishing based on the Peano-like path

References

1. Mizugaki Y, Sakamoto M, Sata T (1992) Fractal path generation for a metal-mold polishing robot system and its evaluation by the operability. CIRP Ann Manuf Technol 41(1):531–534
2. Tam H-y, Lui C-h, Mok ACK (1999) Robotic polishing of free-form surfaces using scanning paths. J Mater Process Technol 95(1):191–200
3. Tam H-y, Lui C-h (1999) Toward the uniform coverage of surfaces by scanning curves. Comput Aided Des 31(1):585–596
4. Cho U, Eom D, Lee D, Park J (1992) A flexible polishing robot system for die and mould. In: Proceedings of the 23rd international symposium on industrial robots, Barcelona, Spain, pp 449–456
5. Pessoles X, Tournier C (2009) Automatic polishing process of plastic injection molds on a 5-axis milling center. J Mater Process Technol 209(7):3665–3673
6. Dunn CR, Walker DD (2008) Pseudo-random tool paths for CNC sub-aperture polishing and other applications. Opt Express 16(23):18942–18949
7. Hu H, Dai Y, Peng X (2010) Restraint of tool path ripple based on surface error distribution and process parameters in deterministic finishing. Opt Express 18(22):22973–22981
8. Deng W-j, Zhang X-j, Wang X-k, Wang X (2009) Novel method for optimizing polishing tool-path in CCOS based on weighted-iterative algorithm. In: 4th international symposium on advanced optical manufacturing and testing technologies: advanced optical manufacturing technologies, International Society for Optics and Photonics, pp 728214–728216
9. Zhang L, Tam H, Yuan C, Chen Y, Zhou Z (2002) An investigation of material removal in polishing with fixed abrasives. Proc Inst Mech Eng Part B J Eng Manuf 216(1):103–112
10. Zhang L, Tam H, Yuan C, Chen Y, Zhou Z, Zheng L (2002) On the removal of material along a polishing path by fixed abrasives. Proc Inst Mech Eng Part B J Eng Manuf 216(9):1217–1225

Chapter 4
Dwell-Time Map

Abstract This chapter focuses on the solution of the dwell-time map, which involves traditional convolution iteration and Fourier transform methods and the sub-aperture corrosion iteration method. Four compensation models for the implementation errors of the dwell-time map are also investigated to further improve fabrication efficiency.

As the convolution model of material removal indicates, if the tool influence function (TIF) has been confirmed to be an invariant parameter, the dwell-time map would control the actual magnitude of material removal. Thus, the calculation of dwell time is a key problem in computer-controlled deterministic optical surfacing systems. Many researchers have undertaken a lot of work in past decades, and Jones [1] proposed a model that utilizes the iterative convolution method to solve the dwell-time map. Wilson and McNeil [2] translated the convolution into frequency domain by Fourier transform. They obtained the dwell-time map through inverse Fourier transform of the division between the Fourier transform of the surface error map and TIF. Carnal et al. [3] transferred the discrete convolution model to a linear equation model and obtained the dwell-time map through the least-square QR decomposition (LSQR) method. It can simultaneously take the shape error and practical polishing path into a fabrication matrix (with rows equal to the sum of total points of the error map and polishing path and columns equal to the number of polishing path points). The dwell time can be obtained by solving a linear ill-posed equation, but nonetheless, no precise solution exists. Researchers have to search a least-squares solution to guarantee a fine convergence rate of the surface error map while avoiding the appearance of negative dwell time.

H. Cheng, *Independent Variables for Optical Surfacing Systems: Synthesis, Characterization and Application*, DOI 10.1007/978-3-642-45355-7_4,

4.1 Calculating Dwell Time by the Convolution Iteration Method

4.1.1 Convolution Iteration Method

Originally, Computer controlled polishing (CCP) operated with dwell time functions proportional to the desired material removal. That was unreasonable when the tool dwelling in the edge region and the convergence rate were unsatisfactory. In 1977, Jones [1] proposed the convolution iteration method. The residual error of a surface form is given by the difference of desired removal ($e_0(m_1 \times m_1)$) and actual removal, which can be described as the convolution of the dwell time function (T) and TIFs ($\mathrm{TIF}(m_2 \times m_2)$).

$$e_i = e_0 - T \otimes \mathrm{TIF} \tag{4.1}$$

Assuming the semidiameter of a workpiece is r_1 and the coverage radius of the TIF is r_2, then m_1, m_2 should satisfy:

$$\frac{r_2}{r_1} = \frac{m_2}{m_1}. \tag{4.2}$$

The dwell-time function is determined by an iterative technique and is initially assumed to be equal to the desired material removal.

- Define the dwell-time matrix $T = 0$; define the iteration time and initial it as $k = 1$.
- Assume the surface form in the kth iteration is e_k, then the dwell time in the kth iteration is $t = e_k/\max(\mathrm{TIF})$, then $T = T + t$; set $k = k + 1$, $e_k = e_0 - \mathrm{TIF} \otimes T$.
- If $\mathrm{RMS}(e_k < e_{k-1})$, then repeat the last step or else stop the iteration and get the dwell-time matrix and residual-error matrix.

To validate the effectiveness of the convolution iteration method used in the calculation of dwell time, a simulation process was conducted and is described in the following. Figure 4.1a represents a surface form (size 400 × 400) related to a Φ100-mm aperture workpiece, which indicates PV = 7.32 λ and RMS = 1.82 λ. A 10-mm aperture TIF (size 40 × 40) was used to figure the surface form, which has a peak removal rate (PRR) of 1.0 λ/min, as shown in Fig. 4.1b. The convolution iteration method then took 12 s to finish the calculation process and produce the dwell-time map, as shown in Fig. 4.1c; the distribution of dwell time is associated with the surface form. The total fabrication time (FT) is 1.77 h. The residual error map is presented in Fig. 4.1d, which shows PV = 0.081 λ and RMS = 0.00345 λ. The convergence rate of the root mean square (RMS) reaches:

$$\mathrm{CR} = \frac{1.82 - 0.00345}{1.82} = 99.82\ \% \tag{4.3}$$

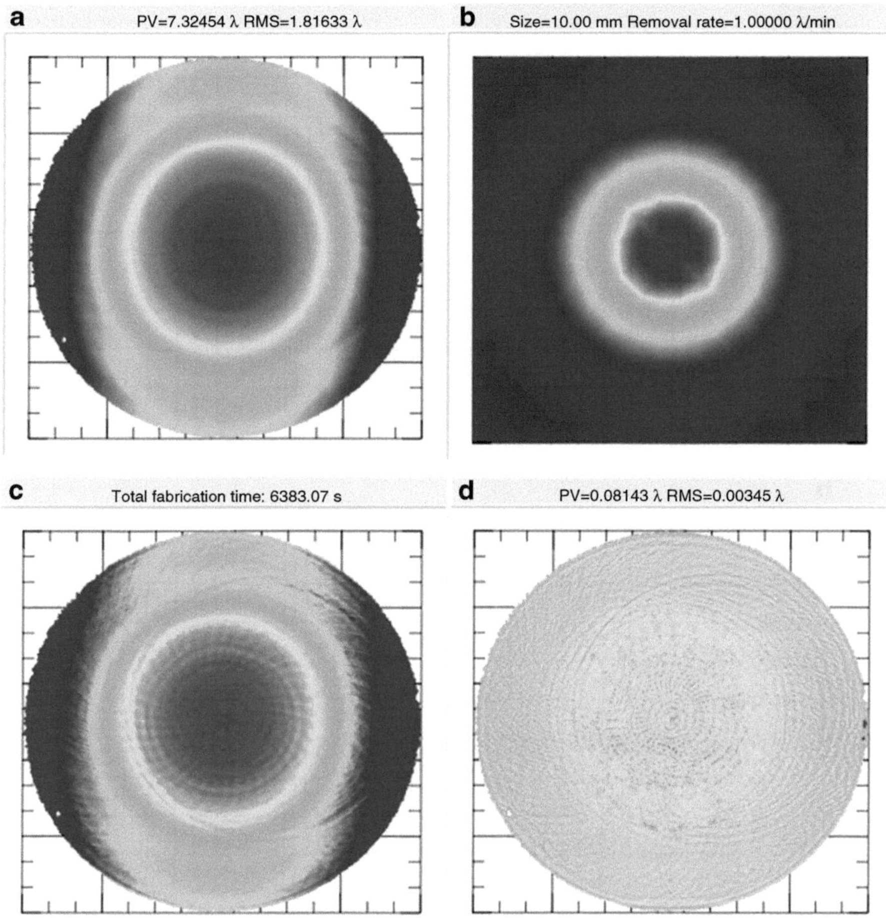

Fig. 4.1 Simulation results and process: (**a**) initial surface form related to a Φ100 mm workpiece; (**b**) TIF used in simulation process; (**c**) dwell-time map; (**d**) residual error map, PV = 0.0814 λ, RMS = 0.0035 λ. *TIF* tool influence function

Figure 4.2a, b are the power spectral density curves before and after the simulation process along the *X* and *Y* directions. The magnitude was reduced more than three decades, and newly produced mid-spatial frequency errors did not appear, which is significant for deterministic fabrication.

Figure 4.3a, b are the slope error maps before and after simulations. Before the simulation process, lots of large slope errors are located in the surface, which may correlate with the grinding mask. After simulation, the slope errors are largely suppressed and removed. The peak-to-valley (PV) of the slope is reduced from 295.4442 to 53.2239 μrad, and the RMS of the slope is reduced from 100.8533 to 6.4075 μrad.

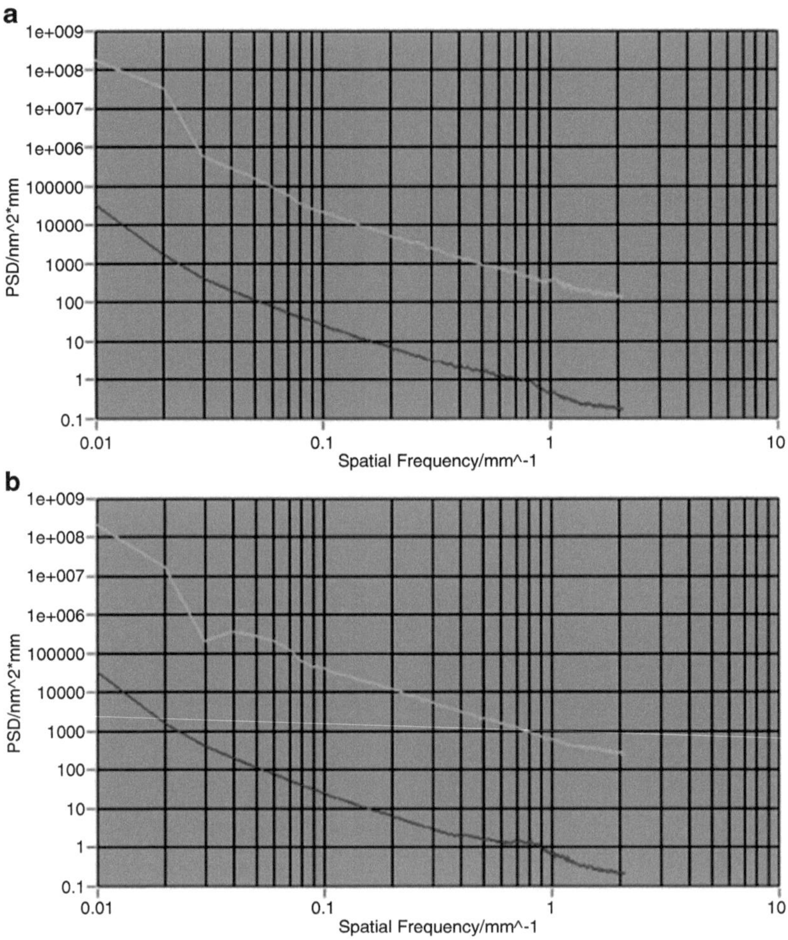

Fig. 4.2 PSD curves before and after the simulation process: (**a**) along *X* direction; (**b**) along *Y* direction. *PSD* power spectral density

4.1.2 Performance Conclusion

The convolution method is frequently used in deterministic polishing of optical mirrors. It has advantages in:

1. The convergence rate is high, which reached 99.82 % in the above simulation, with the slope convergence rate reaching 96 %.
2. The algorithm would not generate newly produced mid-spatial frequency errors. The residual error map is gradual.

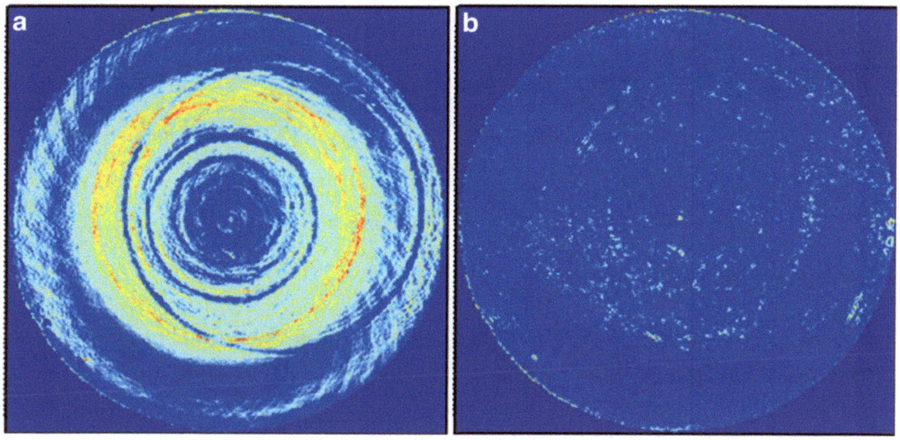

Fig. 4.3 Slope error map (**a**) before and (**b**) after the simulation process

However, the method also needs improvement as follows:

1. The convolution iteration method is based on the two-dimensional (2D) convolution algorithm. Thus, time-wasting is the key problem, especially when the convolution discrete matrix has a large scale. Our test on a common PC shows that it takes more than 10 min to cope with a surface error matrix (scale: $10^3 \times 10^3$) and a TIF (scale: $10^2 \times 10^2$).
2. The simulation process cannot take the polishing path into consideration, which more or less seems to be not suitable to practical fabrication process. To use this algorithm, the dwell time should be allocated to their nearest dwell points. However, this is quite complex and time-consuming.

4.2 Fourier Transform Method

4.2.1 Description of the Algorithm

In the field of optical fabrication, the removal amount d(x, y) is convoluted by the TIF—$r(x, y)$, with a view to the input data (here only consider the dwell time $t(x, y)$). Then d(x, y) is:

$$d(x,y) = kr(x,y) \otimes t(x,y) + e(x,y) \tag{4.4}$$

where k is the removal constant. For convenience, k is combined with $t(x, y)$, that is $g(x, y) = kt(x, y)$. The function $e(x, y)$ is the residual error introduced in the process of optical fabrication. That means the signal $t(x, y)$ is 'polluted' by the noise signal $e(x, y)$. In Eq. (4.4), $d(x, y)$ and $r(x, y)$ are known, and are tested by a profilometer or other methods. However, since signal $t(x, y)$ and signal $e(x, y)$ are unknown, signal $t(x, y)$ cannot be solved just through Eq. (4.4) or, in other words, there are no unique

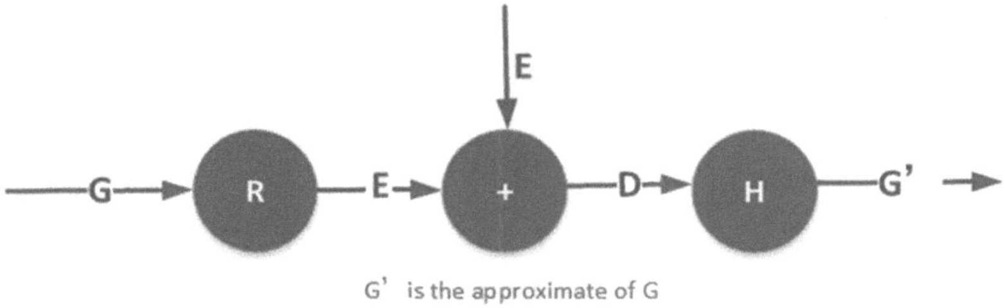

Fig. 4.4 One method of approximation

solutions satisfying Eq. (4.4). Therefore, other restraint conditions have to also be introduced. Converting to the Fourier form, Eq. (4.4) changes into:

$$D(\omega_x, \omega_y) = R(\omega_x, \omega_y) \cdot G(\omega_x, \omega_y) + E(\omega_x, \omega_y) \tag{4.5}$$

where $D(\omega_x, \omega_y)$, $R(\omega_x, \omega_y)$, $G(\omega_x, \omega_y)$, $E(\omega_x, \omega_y)$ represent 2D Fourier translation of signal $d(x, y)$, $r(x, y)$, $g(x, y)$ and $e(x, y)$, respectively. At the next sections, the $D(\omega_x, \omega_y)$, $R(\omega_x, \omega_y)$, $G(\omega_x, \omega_y)$, and $E(\omega_x, \omega_y)$ are stood by D, R, G, E for convenience, despite their dimensions (whether 2D or not, the principle is the same). Therefore:

$$G = \frac{D}{R} - \frac{E}{R} \tag{4.6}$$

If D is not polluted by noise, that is, D is a 'clean' signal D', then G is easily solved by D'/R. When the signal is polluted, at some frequency R is near to zero, but D keeps constant (assume the noise is a 'white' noise), so D/R is not the approximate solution of G, and G cannot be equal to D/R. At this circumstance, the solution of D/R is vibration and divergence; therefore, one of the restraint conditions should be established for solving an approximate solution, which is the least-square approximation or Laplace approximation of G, as shown in Fig. 4.4.

As shown in Fig. 4.4, the signal G passes the system R, then a clean signal D' is obtained. After D' is polluted by E, D is the real signal. G' is the approximation of G, and it can be seen that D is filtrated by H. H is the inverse filter system. The approximation can be least square or smoothly approximate by Laplace's method.

Although getting the exact solution of Eq. (4.6) is not easy, some features of D and E are already known, which will be helpful for our solution. From the Fourier translation, the signal is constituted with all kinds of sine functions at different frequencies. On the condition that the Nyquist frequency twice larger than the signal frequency, the frequency is exact. Usually, at low frequency, the amplitude of D is much larger than that of E, and the amplitude of R is larger than zero, therefore D/R in Eq. (4.6) contributes to the solution, and D/R is close to D'/R. For a large range of frequency, the amplitude of E can be seen as a constant. At high frequency, the amplitude of D attenuates to less than that of E, and the amplitude of R is close to zero; the result can then be determined by E/R. If D is decomposed into two parts, D_1 and D_2 (D_1 is the low frequency of D and D_2 is the high frequency of D), then:

$$D = D_1 + D_2 \tag{4.7}$$

Let D_1/R be approximate to D′/R. That is:

$$G = D^{'}/R \approx D_1/R \tag{4.8}$$

$$E = D_2 \tag{4.9}$$

The meaning of add ('+') in Eq. (4.7) is that D can be divided into two parts: low frequency (D_1) and high frequency (D_2). The meaning of divide ('/') in Eq. (4.8) will be explained later. Equations (4.8) and (4.9) do not give us the unique and exact solution, but give us all the approximate solutions and all the information of optical fabrication that exist in them. In actual work, least-square approximation of G cannot obtain the best solution. Some other conditions, such as the smoothness or oscillation, must be considered together. Equations (4.8) and (4.9) can present a series of solutions that give us a wide range of choice. The next section explains the method of division in Eq. (4.8), because D_1 can be described by Eq. (4.10):

$$D_1 = \sum_{k=0}^{N-1} D(k)\delta(\omega - \omega_k) \tag{4.10}$$

where δ is a delta function, $D(k)$ is the Fourier coefficient of D_1, and R can be described as follows:

$$R = \sum_{k=0}^{\infty} R(k)\delta(\omega - \omega_k) \tag{4.11}$$

Assuming:

$$G = \sum_{k=0}^{N-1} G(k)\delta(\omega - \omega_k) \tag{4.12}$$

where $G(k)$ is the Fourier coefficient of G, therefore:

$$\begin{aligned} R \cdot G &= \sum_{k=0}^{\infty} R(k)\delta(\omega - \omega_k) \sum_{k=0}^{N-1} G(k)\delta(\omega - \omega_k) \\ &= \sum_{k=0}^{\infty} R(k)\delta(\omega - \omega_k) \sum_{j=0}^{N-1} G(j)\delta(\omega - \omega_j) \\ &= \sum_{k=0}^{\infty} R(k) \sum_{j=0}^{N-1} \delta(\omega - \omega_k) G(j)\delta(\omega - \omega_j) \\ &= \sum_{k=0}^{N-1} R(k)G(k)\delta(\omega - \omega_k) \end{aligned} \tag{4.13}$$

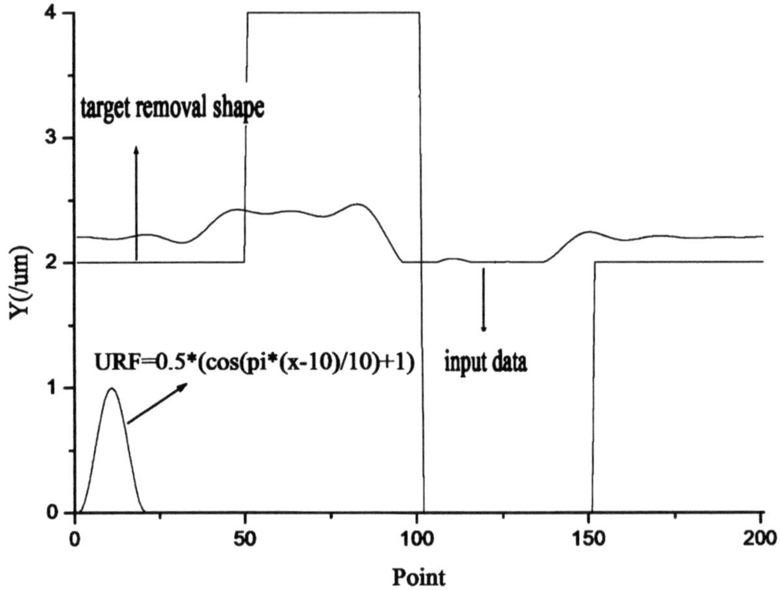

Fig. 4.5 Basic relationship for solving input data

Comparing Eq. (4.10) with Eq. (4.7), get:

$$D(k) = R(k)G(k) \tag{4.14}$$

$$G(k) = D(k)/R(k) \tag{4.15}$$

The input data can be solved by Eq. (4.15). Defining a k dimension space $\phi = \{k_i | i = 1, 2 \ldots n\}$, here n stands for the approximate item (AI). The accuracy is determined by Eq. (4.9), therefore, when n is large, too much of D is removed and accuracy can be improved greatly. However, too much noise will also be enlarged simultaneously by Eq. (4.15), which results in the oscillation of the input data and cannot be implemented by the actual working. Therefore, a reasonable range of n must be chosen. In the next section, the input data with n from 1 to a large number is justified, and some parameters are put forward to evaluate the characteristics of the input data.

From the description above, at a certain range, D_1/R is approximate to D'/R. That is, the range of parameter of n is limited, otherwise, if n is too large, the error will be enlarged at the same time, and the input data G will be oscillated. Therefore, a good range of n can be settled down from 1 to a large number by judging the characteristics of the input data, here, some parameters are defined to judge the effects of the input data.

Figure 4.5 is a curve of TIF (R), target removal shape (D), and one solution of the input data (G, $n = 10$) according to Eq. (4.15). The input data line is displaced two

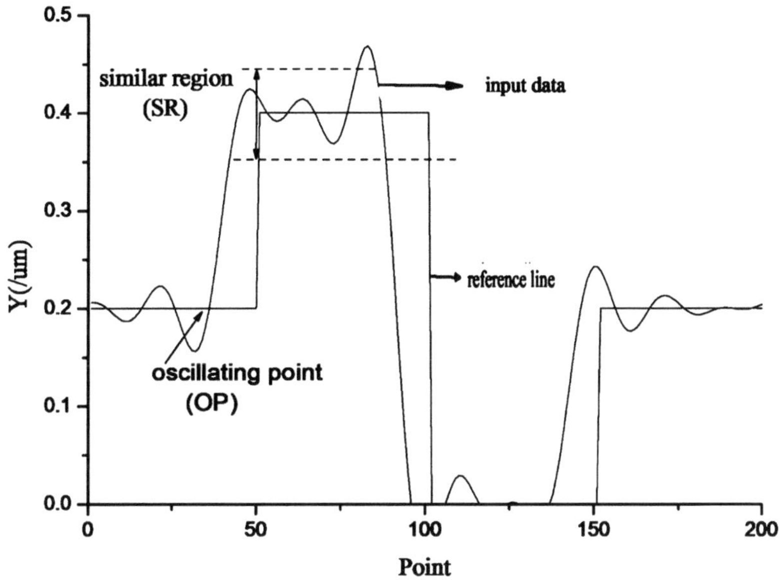

Fig. 4.6 Detailed curves of basic relationship for solving input data

units. During actual manufacturing, the residual errors of the workpiece surface must be removed gradually to achieve a relatively stable error convergent accuracy. In other words, the target removal rate should be divided into several levels referring to the removal quantity of unit removal function (URF). Here, defining $Q = \int_{-\infty}^{+\infty} \mathrm{TIF}(x)\mathrm{d}x = \sum_{n=0}^{N-1} \mathrm{TIF}_n$, Fig. 4.6 is a detailed curve of basic relationship for solving input data shown in Fig. 4.5; the reference line is target removal shape divided by Q.

Some parameters are used to evaluate the effect of algorithm.

RMS. Meaning how much data deviates from the mean value, including RMS of residual errors and that of the input data.

Similar Coefficient (SC). Showing the input data deviate from the reference line along with the AI n. The input data should be at certain arrange. As shown in Fig. 4.6, drawing two lines, go along the reference line with the same deviation, i.e. const $= \|d - d_{\mathrm{ref}}\|$. The region between the two lines is named the 'similar region' (SR), which is used for reflecting how much the input data resemble the reference line. If one input datum drops into the region, counter increments by 1. Then counter divided by datum number which is on the reference line and defined with SC. This curve reflects the degree of similarity between the input data with the reference line; the closer the SC is to 1, the better the result is.

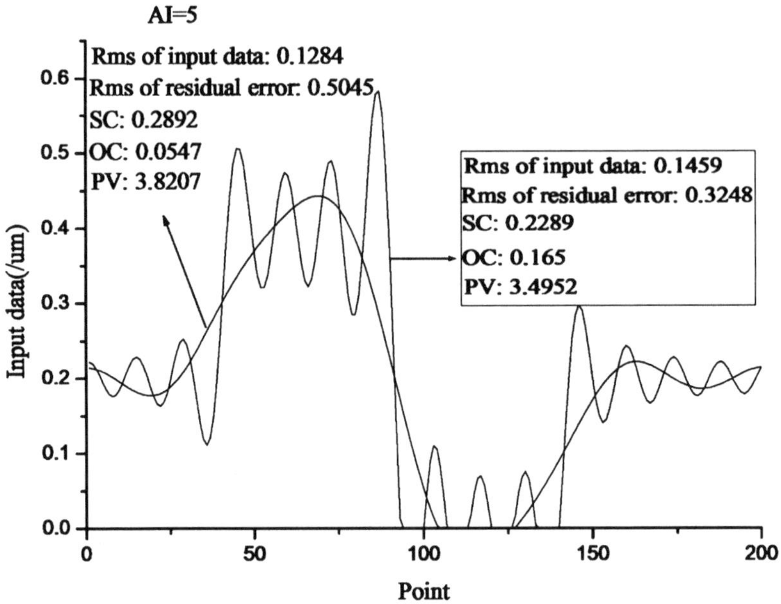

Fig. 4.7 Different AI with different pattern. *AI* approximate item

Oscillating Coefficient (OC). OC reflects the oscillating features of the input data. The oscillating counter is increased by 1 per time the input data crosses over its fair line. The counter divided by data number on the reference line defined with another parameter. OC increases along with the increase of AI used for evaluating residual error. In optical fabrication process, low OC is advisable. However, one point should be noted, OC of the input data is incompatible with that of the residual error.

PV of Residual Errors. PV reflects the largest remaining error amplitude.

Since the effect of the algorithm can not be evaluated fully only by one single parameter, the above parameters reflect the characteristics of the algorithm from different sides. Figure 4.7 is a curve of two sets of the input data at two different choices of AI (AI = 5 and AI = 15). In this figure, some intuitionistic features of these parameters are shown.

4.2.2 Results with Different Approximate Item

Figure 4.8a–f show results obtained from the use of this algorithm. Each figure stands for the trend of a parameter with AI, with n changing from 1 to 30.

Figure 4.8a shows that RMS of residual errors decreases continually until reaching the bottom, along with the increasing AI n from 1 to 18. It then increases sharply with AI n increasing. At the same time, the RMS of dwell time (shown in Fig. 4.8b)

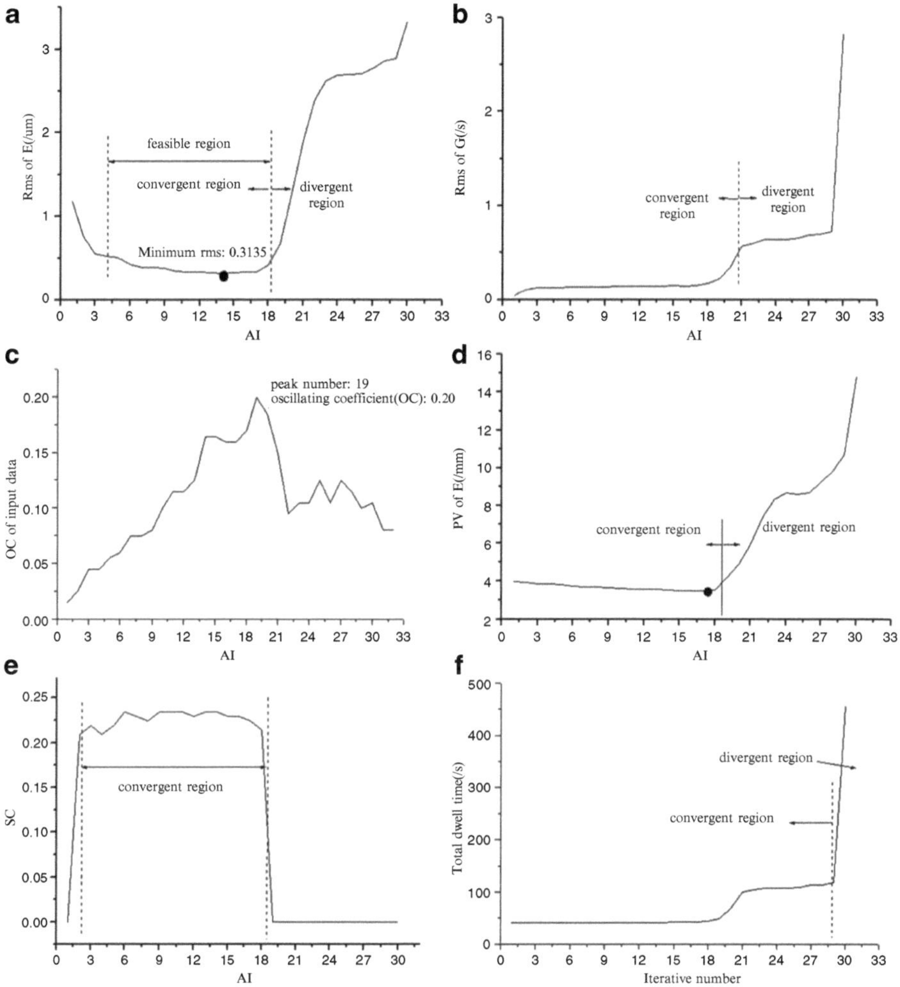

Fig. 4.8 Results from the use of this algorithm. (**a**) Relationship between RMS of residual error with AI; (**b**) relationship between RMS of dwell time with AI; (**c**) relationship between OC with AI; (**d**) relationship between PV of residual error with AI; (**e**) relationship between SC with AI; (**f**) relationship between total dwell time with AI. *RMS* root mean square, *AI* approximate item, *OC* oscillating coefficient, *PV* peak-to-valley, *SC* similar coefficient

increases directly, especially as there is a platform, with AI n increasing from 21 to 29. The two figures convey the information that the lower AI n, the better the result is. However, a key question is still left for determining the exact number of AI n.

With AI n increasing, OC (input data) increases and reaches its highest peak when AI n equals 19 (see Fig. 4.8c), then drops sharply, which means that with AI increasing, the input data vibrates increasingly and makes the accurate fabrication

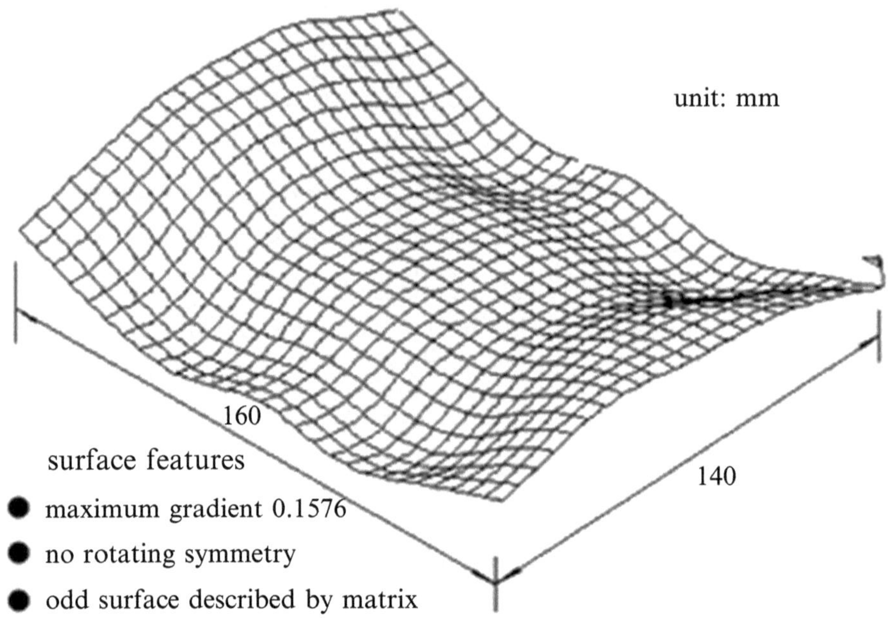

Fig. 4.9 Mesh figure of an example free-form lens FFL

for free-form lens (FFL) become very difficult. PV of residual errors keeps level, then severely increases (see Fig. 4.8d).

As Fig. 4.8e shows, SC has a better region with AI *n*. A clear trend shows that the result has a better feasible region, although it is far away from 1. From the last figure (Fig. 4.8f) for illustrating the situation of total input data (at our case, the input data is dwell time), a conclusion can be made that the result is feasible when AI is smaller than 29.

Figure 4.9 is an example of an FFL. From this figure we can see that the FFL is an odd, complex, warping surface that cannot be described by a continuous function but by discrete data.

Adopting computer-controlled optical surfacing (CCOS) technique to fabricate FFL is effective. During the fabricating process, the surface is first ground by a bound abrasive wheel, then lapped with small diamond particles, and then finally polished. If the error after these fabrication steps is too large, additional error correction has to be performed. Table 4.1 shows parts of the error data of one FFL, which was obtained from a profilometer with a measuring precision of 1 μm; sampling step is 6 mm, and represents the difference between the tested data and the design data. Its unit is micron.

Table 4.2 is the measuring results after 90 h grinding and pre-polishing; the sampling step and unit are the same as those in Table 4.1. Contrast Tables 4.1 and 4.2, accuracy has been improved greatly; maximum error value decreased to 4 μm after error correction from that of 23 μm before error correction. RMS value of residual errors decreased to 0.3 μm.

Table 4.1 Original error data of an FFL in microns

	X/mm					
Y/mm	0	6	12	18	24	30
0	−5	6	6	7	8	11
6	4	6	6	6	7	9
12	2	3	−3	6	7	9
18	1	3	0	6	8	13
24	2	−2	6	10	12	16
30	6	8	12	14	17	23

Table 4.2 Residual error data after grinding and pre-polishing for 90 h refer to the error correction model in microns

	X/mm					
Y/mm	0	6	12	18	24	30
0	2	4	3	3	2	3
6	1	4	3	−1	3	2
12	3	3	3	2	3	2
18	−1	2	0	4	3	4
24	0	4	4	3	4	3
30	4	2	2	3	4	4

Table 4.3 Residual error data after fine polishing for 110 h refer to the error correction model in microns

	X/mm					
Y/mm	0	6	12	18	24	30
0	0.0	0.6	0.9	0.4	0.6	0.5
6	0.1	0.1	0.6	0.2	0.4	0.9
12	0.6	0.4	0.5	0.7	0.1	0.3
18	0.5	0.2	0.7	0.8	0.9	0.5
24	0.9	0.7	0.3	0.6	0.4	0.2
30	0.5	0.9	0.2	0.4	0.3	0.7

To further improve the manufacturing accuracy of the FFL, kept at AI $= 15$, we continued to fabricate the workpiece based on the established error correction model referring to the data achieved by using the non-contact optical measuring method with measuring precision of 0.1 μm. Table 4.3 is part of the measuring data of the workpiece surface after 110 h fine polishing. Figure 4.10 is the error convergent curve in different machining stages based on the error correction model. PV value of residual errors reduced to 0.9 μm and RMS value decreased to 0.097 μm. Here, sampling points presented in Tables 4.1, 4.2, and 4.3 have the same coordinate position (X_i, $Y_j|i$, $j = 0, 1, \ldots 5$).

Figure 4.11a–d are the fitting three-dimensional (3D) mesh graph of the original error data, error data after 90 h grinding and pre-polishing, error data after 110 h fine polishing, and the photograph of the finished FFL, respectively. These figures prove the finite Fourier coefficient algorithm is appropriate for the error correction of FFL, and the actual shape can be expressed accurately by the input data after error correction.

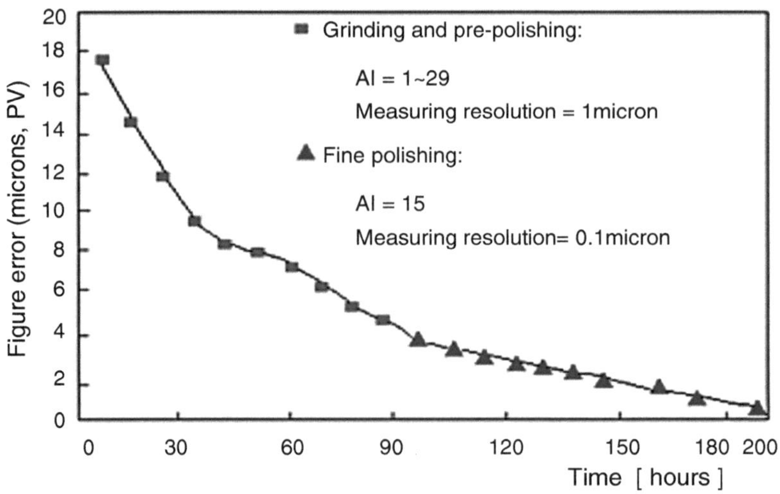

Fig. 4.10 Figure error convergent curve

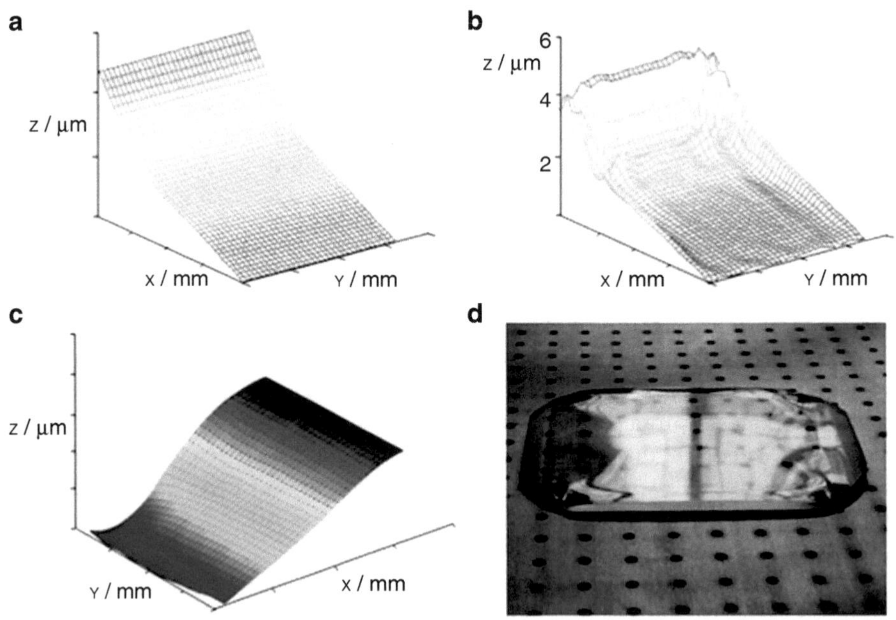

Fig. 4.11 3D mesh figure and photograph of the FFL. (**a**) Distribution view of original error data; (**b**) distribution view of error data after 90 h pre-machining; (**c**) distribution view of error data after 110 h fine polishing; (**d**) photograph of the finished FFL. *3D* three-dimensional

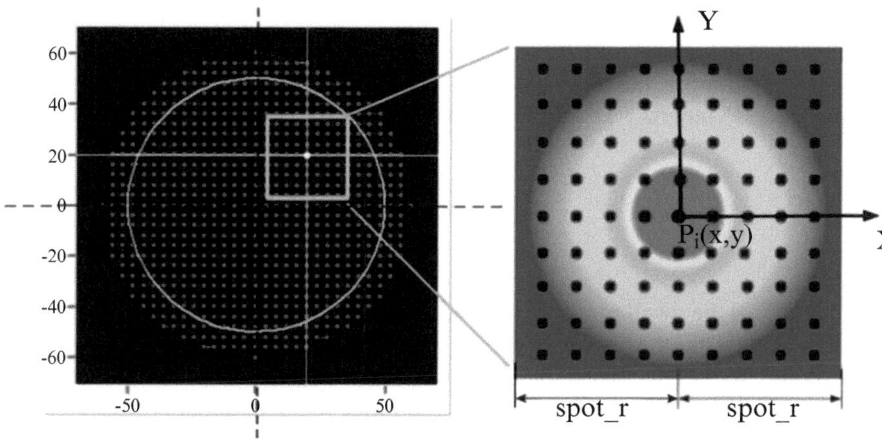

Fig. 4.12 Sketch map of the subaperture corrosion iteration (SACI) dwell-time algorithm

4.3 Subaperture Corrosion Method

4.3.1 SACI Dwell-Time Algorithm

As the algorithm is an optimized iteration method, an object merit function should be predefined to decide the conditions for stopping iteration. It should simultaneously take the low-spatial and mid-spatial frequency errors into account. The object merit function is defined as Eq. (4.16).

$$\mathrm{OMF} = \left(C_1 \cdot \mathrm{RMS}^2 + C_2 \cdot \mathrm{Slope\_RMS}^2\right)^{0.5} \tag{4.16}$$

In Eq. (4.16), OMF denotes the object merit function; the RMS represents low-spatial error, and Slope_RMS represents the RMS of slope data which is utilized to describe the mid-spatial frequency errors. The coefficients C_1 and C_2 represent the weight factors of low-spatial frequency RMS and slope RMS, respectively. If we only care about the RMS, then C_1 should be set as 1 and C_2 should be set as 0.

In fact, the sub-aperture corrosion is an imitation of practical polishing process. According to the location of dwell point, take out a submatrix with the same scale as the TIF from the surface error matrix, and subtract the arithmetic product of the TIF and its dwell time coefficient. Then, by accumulating corrosions in the spatial polishing path, the target material removal is achieved, where the TIF plays a role of corrodent to figure the surface error.

Assign that: the original shape error matrix e_0, with a scale of $m_1 \times m_1$. The scale of the TIF matrix is $m_2 \times m_2$. Generally, the scales between TIF and e_0 have relationship as Eq. (4.17); where, glass_r and spot_r represent the semi-diameter of workpieces and TIFs, respectively. In Fig. 4.12, the total number of dwell points is N. For the ith dwell point $P_i(x, y)$, its influence area (IA_i) is plotted in the right square with side length L_s. IA_i has the same size with the TIF. In the right square, the background picture is the top view of a typical TIF and the black points are dwell points. The side length L_s could be expressed as Eq. (4.18).

$$\frac{m_2}{m_1} = \frac{\text{spot\_r}}{\text{glass\_r}} \tag{4.17}$$

$$L_s = 2 \cdot \text{spot\_r} \tag{4.18}$$

Circulate all the dwell points $P(x, y)$, and count the number of dwell points in IA_i, assigned as n_i. That means while the polishing tool's center dwelling at any other dwell points in IA_i, it always has removal effect to $P_i(x, y)$. The total removal quantity of $P_i(x, y)$: r_i is the summation of n_i times removal quantity, expressed as Eq. (4.19), where $t(x_j, y_j)$ indicates the dwell time in $P_j(x, y)$.

$$r_i = \sum_{j=1}^{n_i} \text{TIF}\left(x_i - x_j, y_i - y_j\right) \cdot t\left(x_j, y_j\right) \tag{4.19}$$

We employ a parameter to evaluate the influence of the polishing path points' density: path weight factor (PWF). The PWF of the ith dwell point (PWF_i) can be expressed as Eq. (4.20). Normalized PFW (NPWF) is presented in Eq. (4.21) as the division of the ith and maximum PWF. It implies, if we want to achieve average removal on a surface, the point has larger normalized PWF and will need more dwell time because of the lower polishing path's density.

$$\text{PWF}_i = \frac{1}{\sum_{j=1}^{n_i} \text{TIF}\left(x_i - x_j, y_i - y_j\right)} \tag{4.20}$$

$$\text{NPWF}_i = \frac{\text{PWF}_i}{\max(\text{PWF})} \tag{4.21}$$

Define convergence speed factor: μ, which could be described as a ratio factor between the removal quantity in the kth iteration process (assigned as $e_k^{'}(x_i, y_i)\mathfrak{M}$) and residual error magnitude after ($k - 1$)th iteration $e_{(k-1)}(x_i, y_i)$ of $P_i(x, y)$, as shown in Eq. (4.22). The convergence speed factor varies as the position of dwell points. For a uniform path, the convergence speed factors of inner dwell points are generally equal. However, if the dwell point is close to the edge, the convergence speed factor increases as the radial distance expanding. For the helix path in Fig. 4.13, the ratio factors on the path points are plotted in Fig. 4.13.

$$\mu_i = \frac{e_k^{'}(x_i, y_i)}{e_{(k-1)}(x_i, y_i)} = \frac{\text{NPWF}_i}{\max(n_i)} \tag{4.22}$$

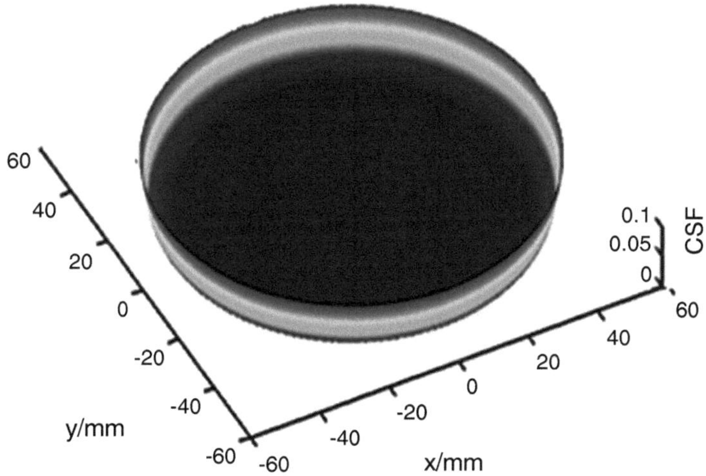

Fig. 4.13 Distribution of convergence speed factors (CSFs) for a helix polishing path

Equation (4.22) gives the empirical value of convergence speed factor. It can control the removal quantity in a single iteration. Generally, a large convergence speed factor can bring higher convergence speed in a single iteration, but if we set it much larger than Eq. (4.22), the TIF would generate excessive material removal, and the residual error after the current iteration may have a mass of negative data, which is unexpected. Therefore, we generally could take the Eq. (4.22) as the practical convergence speed factor. Moreover, it can be seen from Eq. (4.22) that a larger convergence speed factor means it needs more dwell time for the same removal magnitude.

After getting the convergence speed factor, we should confirm the single maximal removal quality of the dwell point of $P_i(x, y)$ in the kth iteration: $e'_k(x_i, y_i)$. Dwell time $t_k(x_i, y_i)$ in $P_i(x, y)$ of the kth iteration can then be obtained. Assume the residual error magnitude of $P_i(x, y)$ after the $(k-1)$th iteration is $e_{(k-1)}(x_i, y_i)$. The dwell time $t_k(x_i, y_i)$ can then be expressed as Eq. (4.23). The residual error map $e_k(x, y)$ can be obtained by Eq. (4.24).

$$t_k(x_i, y_i) = e'_k(x_i, y_i)/\max(\mathrm{TIF}) = \mu_i e_{(k-1)}(x_i, y_i)/\max(\mathrm{TIF}) \tag{4.23}$$

$$e_k(x, y) = e_{(k-1)}(x, y) - \sum_{j=1}^{n_i} \mathrm{TIF}\left(x_j - x, y_j - y\right) \cdot t_k\left(x_j, y_j\right) \tag{4.24}$$

According to the above analysis, the iterative flow sheet of SACI is concluded as Fig. 4.14. The process can be described as follows:

1. Assume the final dwell-time map is $T(x, y)$, initialize the iterative times $k = 0$, and $T(x, y) = 0$.

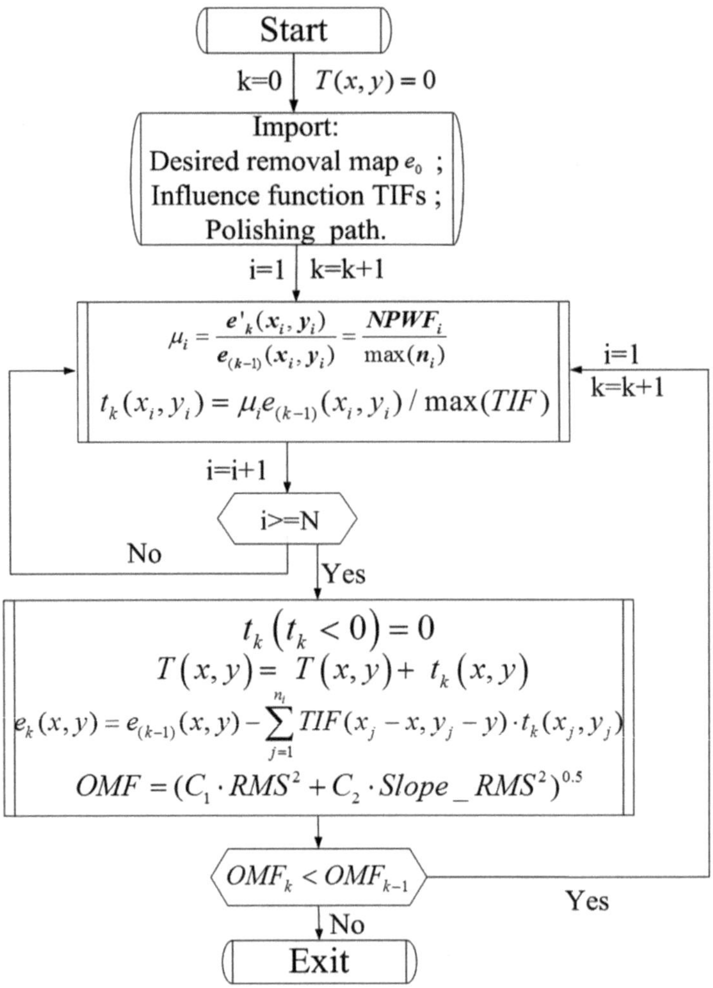

Fig. 4.14 Flow sheet for SACI dwell-time algorithm

2. Input the original surface error map e_0, the TIF used in the polishing run, and the polishing path including X and Y vectors with length of N. Set $k = k + 1$, set initial point's number $i = 1$.
3. Get into the first iteration. Circulate i from 1 to N, obtain $t_k(x, y)$ according to Eqs. (4.22) and (4.23).
4. If negative dwell time appears in $t_k(x, y)$, set the negative data as zero. Then, $T(x, y) = T(x, y) + t_k(x, y)$, and obtain the residual error map after the kth iteration e_k, according to Eq. (4.24).
5. If $\mathrm{OMF}_k < \mathrm{OMF}_{k-1}$, set $k = k + 1$, $i = 1$, get into step (3). Otherwise, stop the iteration and obtain the final dwell time $T(x, y) - t_k(x, y)$ and residual error map e_{k-1}.

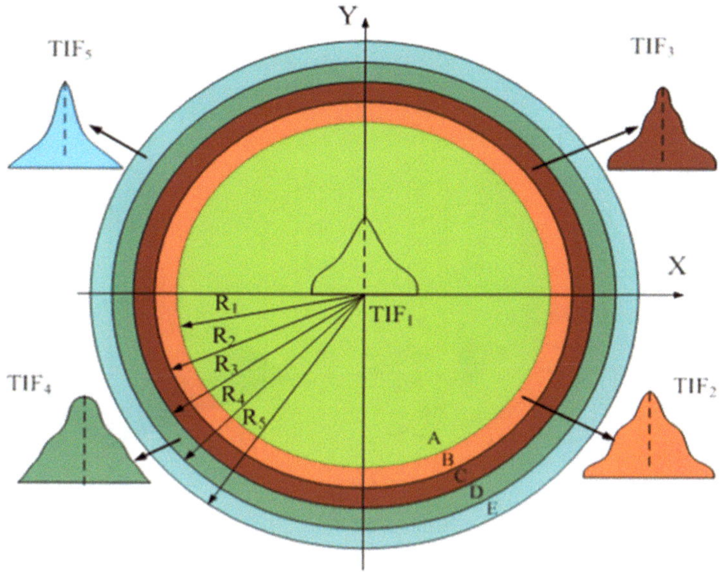

Fig. 4.15 Schematic illustration of multiple TIFs in one optimization. *TIFs* tool influence functions

As the existing edge effect for some polishing setups, the position stability of the TIF could not be guaranteed. Therefore, it is unreasonable if only one invariant TIF is input for a calculation process. The subaperture corrosion iteration (SACI) dwell-time algorithm adopts the numeric calculation method, and is compatible with multiple TIFs in one optimization process. As an example, in Fig. 4.15, the whole workpiece is divided into five portions, assigned as A, B, C, D, E, which correlate with different TIFs. Therefore, five TIFs should be taken into the calculation process, and SACI could select a TIF for every dwell point according to its location. This feature is significant for reducing edge effect.

4.3.2 Simulation Process and Results

A randomly generated surface error map using the Zernike polynomial is shown in Fig. 4.16a. The Zernike coefficient is expressed in Eq. (4.25). The surface error map (scale 400 × 400) correlates with an optical workpiece of Φ200 mm.

$$Z_3 = -1, Z_4 = 1, Z_7 = 1, Z_8 = 1, Z_9 = -1 \tag{4.25}$$

Figure 4.16b is the TIF used in this simulation (aperture 10 × 10 mm). The TIF scans a raster polishing path with pitch 1.0 mm. Both C_1 and C_2 were set as 0.5. After 11 s, the residual error and the dwell-time map were obtained (shown in

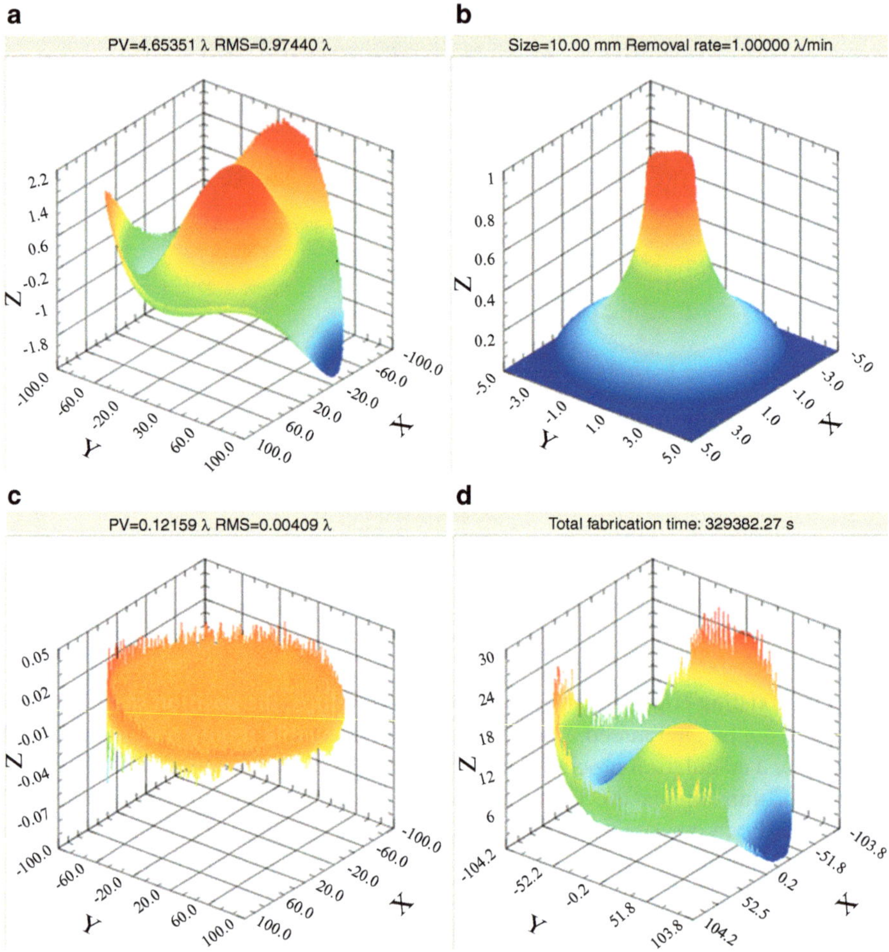

Fig. 4.16 Results of simulation. (**a**) Original error map, PV = 4.6535 λ, RMS = 0.9744 λ, aperture Φ200 mm; (**b**) the used TIF with aperture 10 × 10 mm; (**c**) residual error map in full aperture, PV = 0.1216 λ, RMS = 0.0041 λ, 98 % aperture V = 0.0312 λ, RMS = 0.0021 λ; (**d**) dwell-time map. *PV* peak-to-valley, *RMS* root mean square

Fig. 4.16c, d). PV was reduced from 4.65 λ to 0.12 λ; RMS converged from 0.9744 λ to 0.0041 λ. Few oscillations existed in the edge region, which indicated the edge effect was well suppressed. The dwell-time map is smooth, non-negative, continuous, and one-to-one with the practical polishing path.

Figure 4.17a, b represent the comparison diagrams of the slope error map before and after polishing, respectively. Slope error was obviously improved: Slope_PV converged from 117.99 to 14.678 μrad; Slope_RMS converged from 33.96 to 2.270 μrad.

Figure 4.18a, b are the power spectral density curves before and after polishing along X and Y direction. Every frequency range error was largely decreased, and

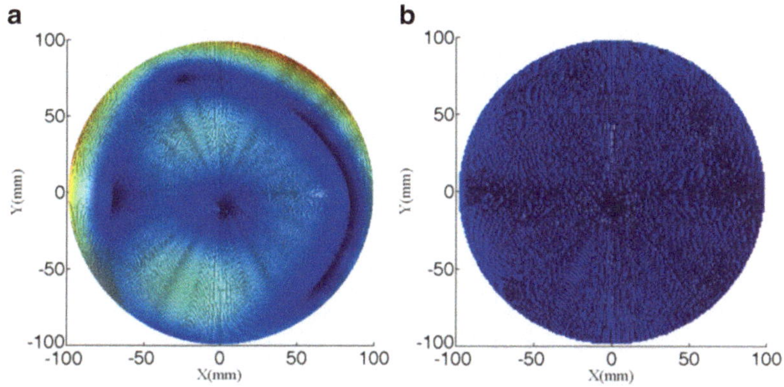

Fig. 4.17 Slope error map of (**a**) before polishing, Slope_PV = 117.99 μrad, Slope_RMS = 33.95 μrad, and (**b**) after polishing, Slope_PV = 14.678 μrad, Slope_RMS = 2.270 μrad. *PV* peak-to-valley, *RMS* root square mean

newly produced mid-spatial frequency errors caused by the polishing tool did not appear, which indicated that SACI has a well-suppressing effect for mid-spatial frequency errors.

From the features of the SACI dwell-time algorithm and the above simulation, the performances of this algorithm are concluded as follows.

1. The polishing paths may be arbitrary and non-uniform. The numbers of path points could also be more or less than the removal points of surface error map.
2. Compared with the matrix-based algorithm, there is no need to adjust the weight factors or other factors to acquire a high convergence rate and non-negative dwell-time map.
3. The dwell-time map is highly smooth and continuous, and negative dwell time does not appear. These characteristics could be convenient for polishing setups implementing the dwell-time map.
4. The high convergence rate is a significant advantage of SACI. In the above simulation, the RMS convergence rate shows: CR_0 reached 99.6 %, as shown in Eq. (4.26); Slope_RMS convergence rate: CR_1 reached 93.3 %, as shown in Eq. (4.27).

$$CR_0 = \frac{0.9744 - 0.0041}{0.9744} = 99.6\ \% \tag{4.26}$$

$$CR_1 = \frac{33.96 - 2.27}{33.96} = 93.3\ \% \tag{4.27}$$

5. Less calculating time is another merit. Four situations were conducted and listed in Table 4.4. Regarding the fourth situation, the surface error has 10^3 rows $\times$ 10^3 columns and the polishing path has 33,020 points; the calculating process only took 22 s on a common PC.

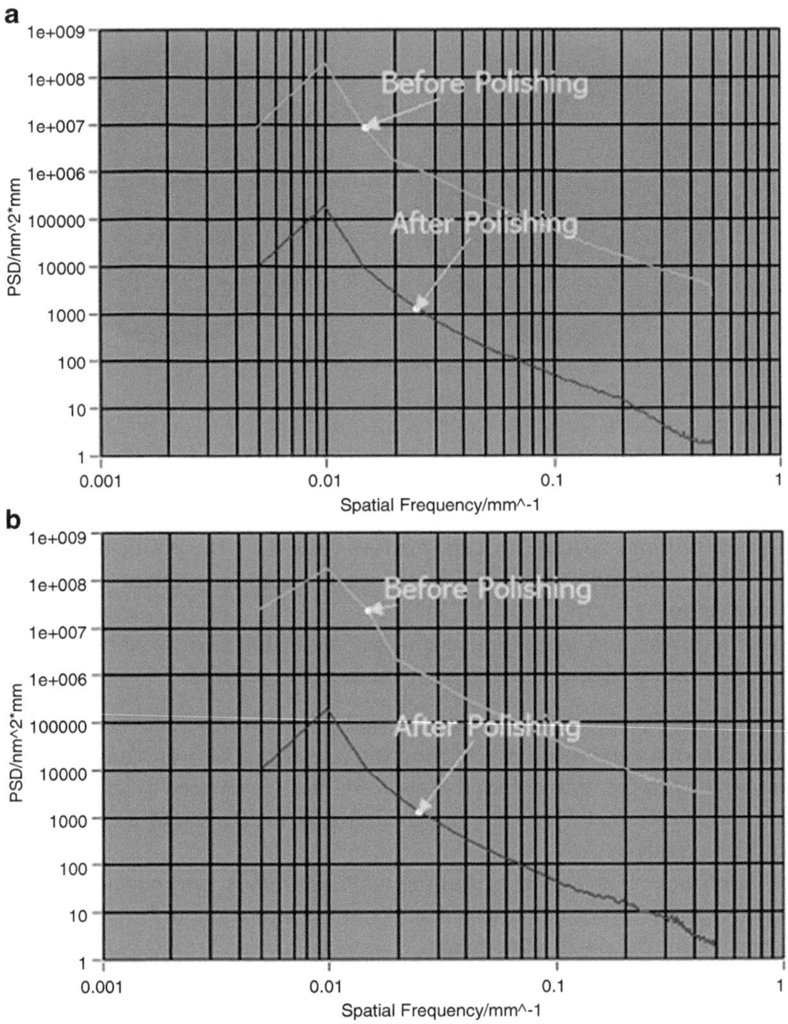

Fig. 4.18 Power spectrum density curves before and after polishing. (**a**) Along *X* direction; (**b**) along *Y* direction

Table 4.4 Calculating time of different size error matrix and polishing path

S/N	Workpiece size (mm)	Surface error map	TIF size (mm)	Path number	Time (s)
1	Φ100	400 × 400	10 × 10	10,080	6
2	Φ200	400 × 400	10 × 10	33,020	11
3	Φ200	800 × 800	10 × 10	33,020	19
4	Φ200	1,000 × 1,000	10 × 10	33,020	22

TIF tool influence function

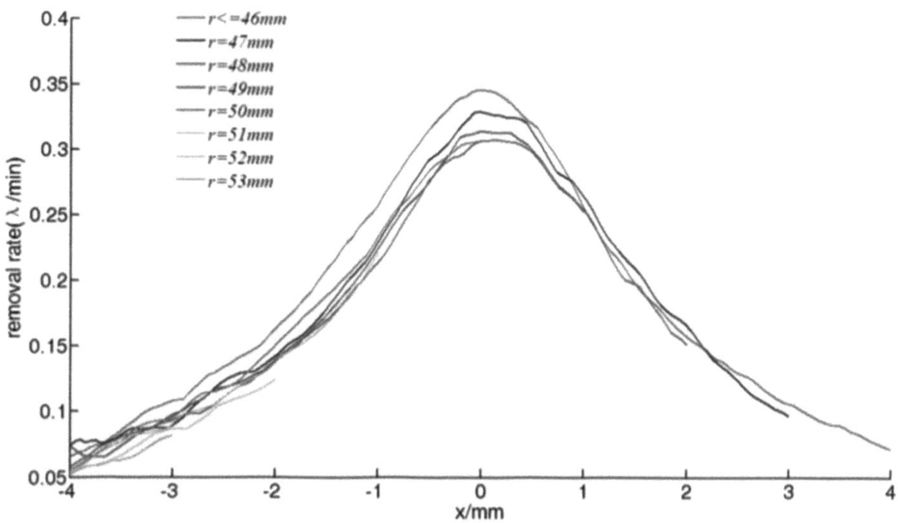

Fig. 4.19 Factual TIFs used in the polishing process, while the polishing tool locates in different regions of the surface. *TIF* tool influence functions

6. Compatible with multiple TIFs in one optimization process. This means that we can input several different TIFs to optimize dwell time for the polishing setups with different edge TIFs, which could help suppress the edge effect.

4.3.3 Validation of Experiments

To validate the effectiveness of SACI in practical fabrication, a Zerodur plane workpiece with aperture of 100 mm was polished on the magnetorheological finishing setup. Cheng et al. [4] designed this machine. It can supply rotational symmetry TIF benefits from its self-rotation and orbital motion model. The TIF has an 8-mm aperture with PRR of 0.346 λ/min; an equal arc length helix path with pitch of 1.0 mm was adopted. Figure 4.19 expresses the factual TIFs in a different region of the surface as we measured. '*r*' denotes the radial distance of the dwell point. As the tool exceeding the edge, its PRR monotonically decreases. These eight TIFs took SACI dwell time into the calculation process. Lastly, we obtained the fabrication file with the dwell-time map, which was subsequently implemented on the surface.

The original surface error map (in full aperture) is shown in Fig. 4.20a: PV = 0.509 λ, RMS = 0.137 λ. After one polishing run, the PV and RMS were reduced to 0.079 λ and 0.013 λ, respectively, as shown in Fig. 4.20b. The edge effect was largely suppressed as the function of extended path. Figure 4.21a, b are power spectrum density curves along X and Y direction before and after polishing, respectively. They indicate the polishing run did not introduce newly produced

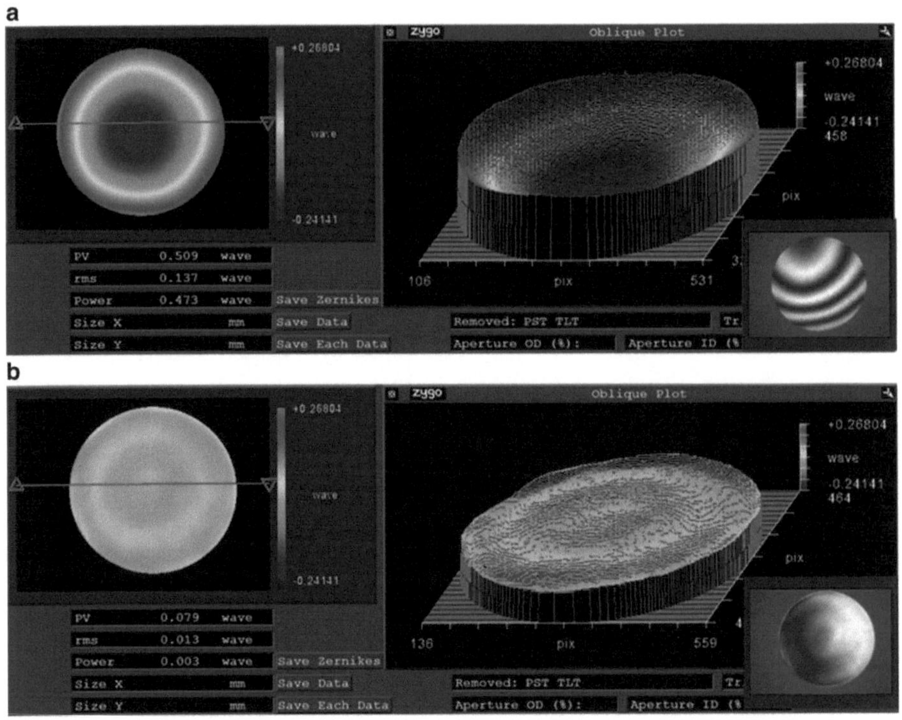

Fig. 4.20 Error maps of (**a**) before polishing: PV = 0.509 λ, RMS = 0.137 λ and (**b**) after polishing: PV = 0.079 λ, RMS = 0.013 λ. *PV* peak-to-valley, *RMS* root square mean

mid-spatial frequency errors. The experiment validated the performance of SACI used in magnetorheological finishing (MRF).

4.4 Implementation of Dwell-Time Map

After the dwell-time map is obtained in the simulation process, the next step is how to accurately implement it on the optical surface. There are two methods that can be used for this implementation: position dwell and speed dwell. The former method dwells the tool on every path's point for a while equal to its desired dwell time. Subsequently, the tool traverses to the next point with the maximum allowed velocity. As shown in Fig. 4.22, the polishing wheel is scanning a raster polishing path with spinning velocity: Vs. It is easy to find that, with the former method, the tool's motion is discontinuous in both the feeding and scanning directions. Thus, this method would introduce more mid-spatial frequency errors due to the foot imprint superposition of TIFs. The trace of the polishing path would be clearly mapped on the surface. Moreover, when the tool is scanning a spiral or concentric

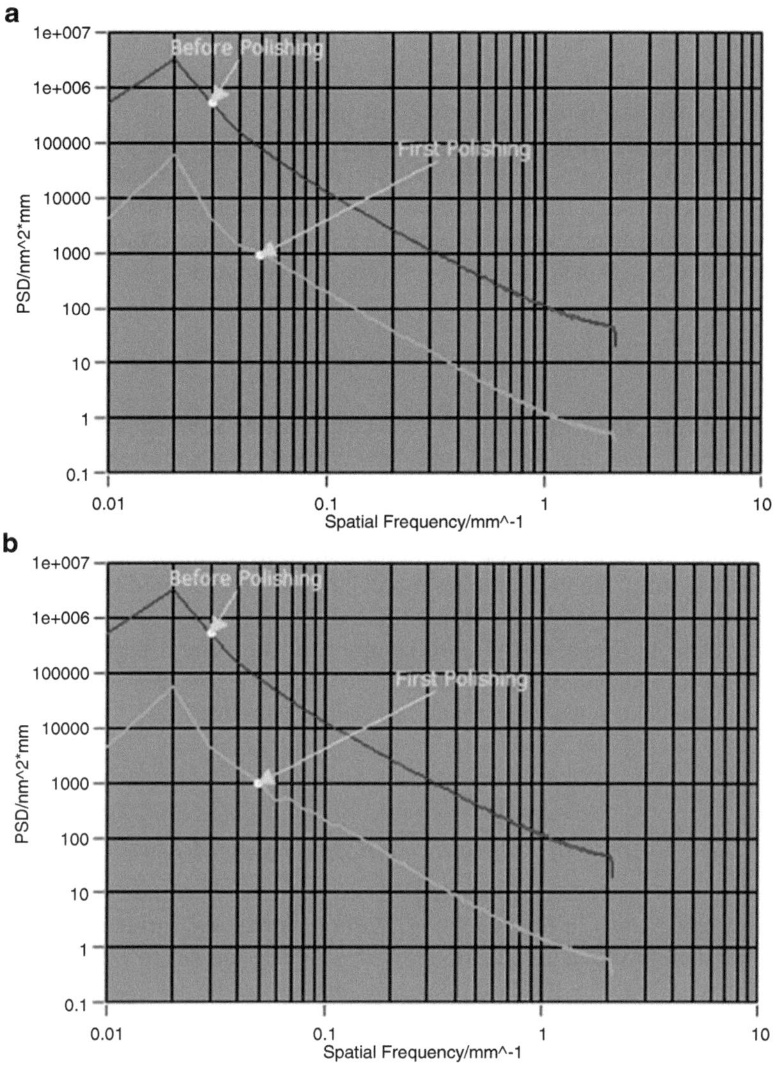

Fig. 4.21 Power spectrum density curves before (*red*) and after (*green*) polishing. (**a**) Along *X* direction; (**b**) along *Y* direction

Fig. 4.22 Schematic diagram of the speed dwell method

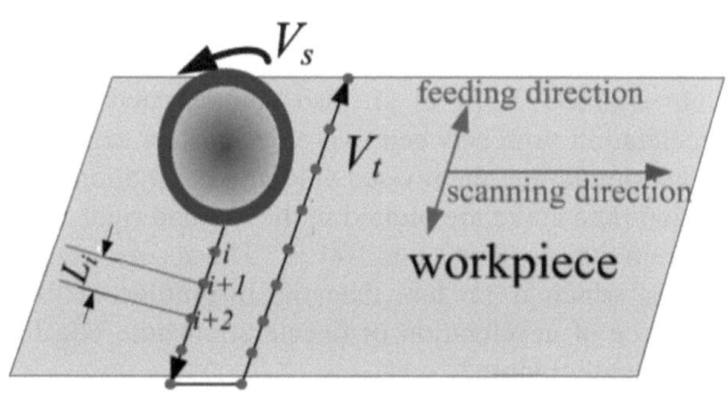

circle polishing path, the motion between two adjacent points inevitably results in non-uniform material removal over the full surface.

The speed dwell method is often adopted by researchers. The polishing machine varies the transverse velocity of the tool or workpieces: V_t. As shown in Fig. 4.22, the line segment which links two adjacent path points can be seen as a basic unit. The desired transverse velocity (V_{ti}) of line segment Li can be directly interpreted from its spatial length and desired dwell time, as presented in Eq. (4.28).

$$V_{ti} = \frac{L_i}{t_i} \tag{4.28}$$

where t_i indicates the dwell time of L_i. Thus, if a line segment correlates with a too-short dwell time, the desired transverse velocity probably exceeds the maximum allowed velocity of the machine: V_m (assumed to be a constant). To protect itself, the machine has to take V_m to transverse through these abnormal points, which would certainly decrease the implementation accuracy of the dwell-time map. Consequently, the tool removes more materials than desired because its dwell time has been prolonged a lot. When the tool scans a spiral or concentric circle polishing path, the problem is more common in the central region because the turntable must rotate much faster than in the edge region (the intersection angle of a unit line in the central region is much larger than the edge).

4.4.1 The Source of Implementation Errors of Dwell Time

As an example shown in Fig. 4.23, Fig. 4.23a indicates the surface error map of a Φ100 mm plane workpiece. The tool produces a Gaussian-shape TIF with Φ10 mm aperture and 1.0 λ/min PRR. If the machine drives the tool scanning a spiral polishing path (with screw pitch = 1.0 mm), the dwell-time map can be obtained by an iterative algorithm, which is shown in Fig. 4.23b. PV and RMS are reduced from 1.96 λ and 0.23 λ to 0.20 λ and 0.021 λ, respectively. The desired velocity distribution map V_t of the turntable is presented in Fig. 4.23c, wherein the maximum value reaches up to 34.810°/s. Considering the maximum velocity of the turntable (e.g. V_m = 10°/s), the revised transverse velocity map V_r is plotted in Fig. 4.23d. The machine will carry out the revised velocity map to implement the dwell-time map. The total FT is 5.97 h.

In addition, Song et al. also compensated the influences of acceleration or deceleration time between two adjacent line segments [5]. For an instance of the above simulation, the revised velocity distribution curve and accelerating or deceleration time curve are plotted in the left and right Y-axis of Fig. 4.24, respectively, wherein the acceleration is 100°/s^2. The accelerating time generally stays less than 0.01 s, which is far less than the magnitude of the dwell-time map. Thus, the influence of acceleration or deceleration time is a less important factor that can be generally ignored.

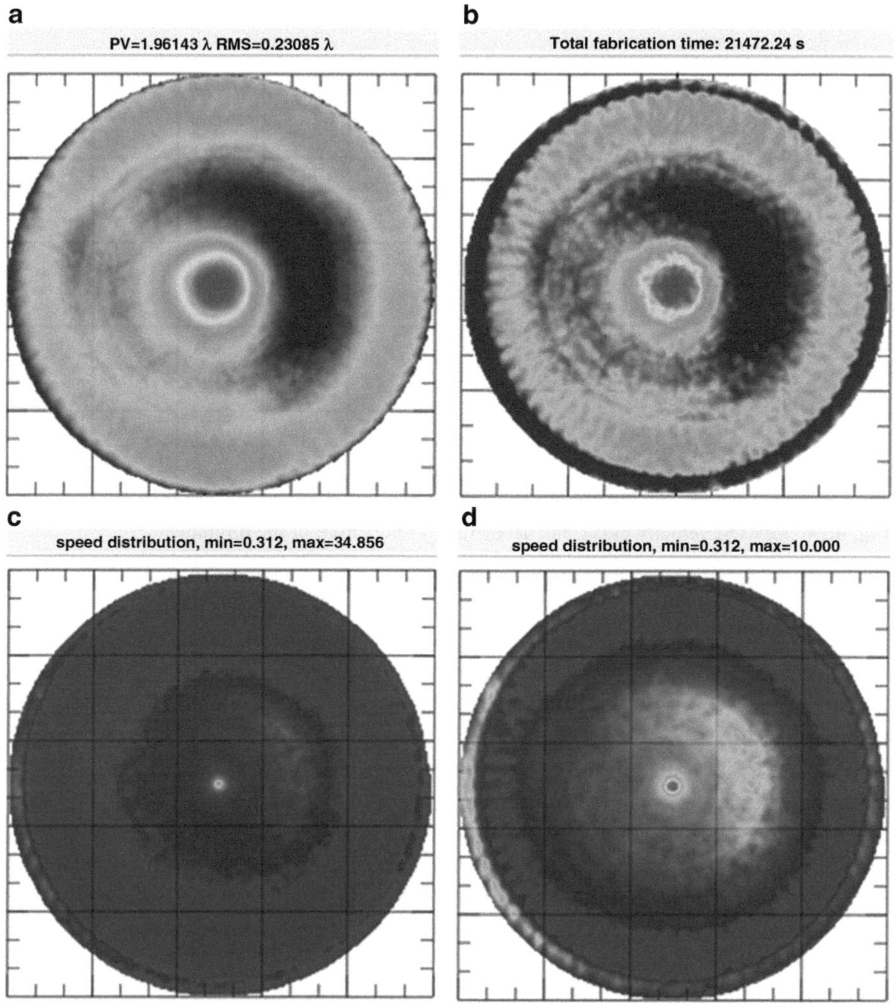

Fig. 4.23 Simulations of: (**a**) original surface error map, PV = 1.96 λ, RMS = 0.23 λ (λ = 632.8 nm); (**b**) desired dwell-time map, 5.97 h in total; (**c**) desired velocity distribution map of the turntable, the maximum value is 34.8°/s; (**d**) revised velocity distribution map, maximum velocity is 10°/s. *PV* peak-to-valley, *RMS* root square mean

4.4.2 Four Compensation Strategies

4.4.2.1 Adding an Extra Fictitious Removal Layer Uniformly on the Surface

Shown as Eq. (4.29):

$$\mathrm{TR}' = \mathrm{TR} + d \tag{4.29}$$

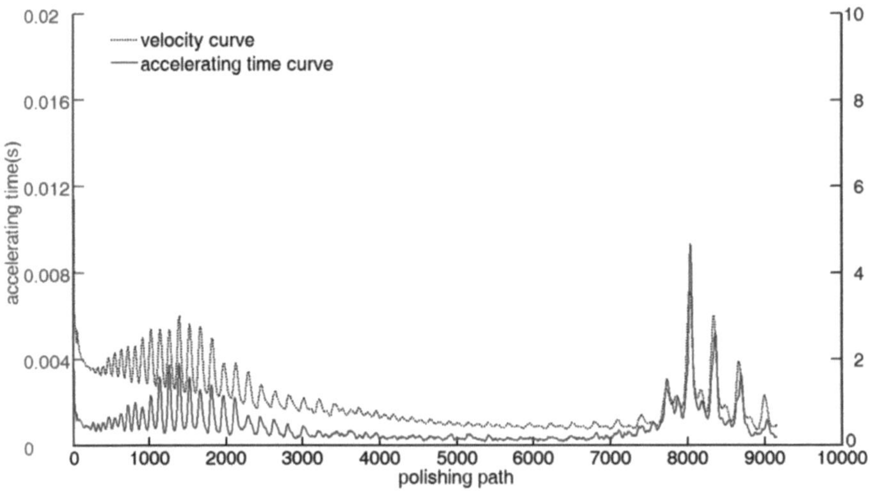

Fig. 4.24 Revised velocity curve and accelerating time curve of the turntable

where "TR" represents the target removal map, which is the deviations of the real surface from an ideal; "TR′"TR indicates the removal map after adding the uniform layer; *d* is the thickness of the adding layer, serving as an additional material removal parameter.

The abnormal points mainly result from a very-short dwell time. Thus, if we add an extra uniform removal layer on the surface, the dwell-time map would be scaled and enlarged. Here, the thickness of the adding layer should be carefully selected to ensure that the desired transverse velocity does not exceed the maximum allowed velocity. However, this method has a strong impact on the total surface. It inevitably increases the total dwell time largely. Thus, it seems more or less unfit to the commercial mass fabrication.

For the simulations as mentioned above, an extra removal layer with a thickness of 6.0 λ should be added to reduce the maximum velocity to less than 10°/s (shown in Fig. 4.25b). The residual surface error map shows PV $= 0.68\ \lambda$ and RMS $= 0.063\ \lambda$. Compared with the simulation without adding layer, the results are enlarged more than three times, and the FT is prolonged to 45.41 h (shown in Fig. 4.25a). These sacrifices for convergence rate and FT are generally unaccepted. Therefore, this method needs constrained condition. It can be considered if the maximum desired velocity (e.g. 12°/s) only slightly exceeds the maximum allowed velocity. Although the method is easy to perform, the convergence rate and fabrication efficiency are both diminished, thereby limiting its versatility.

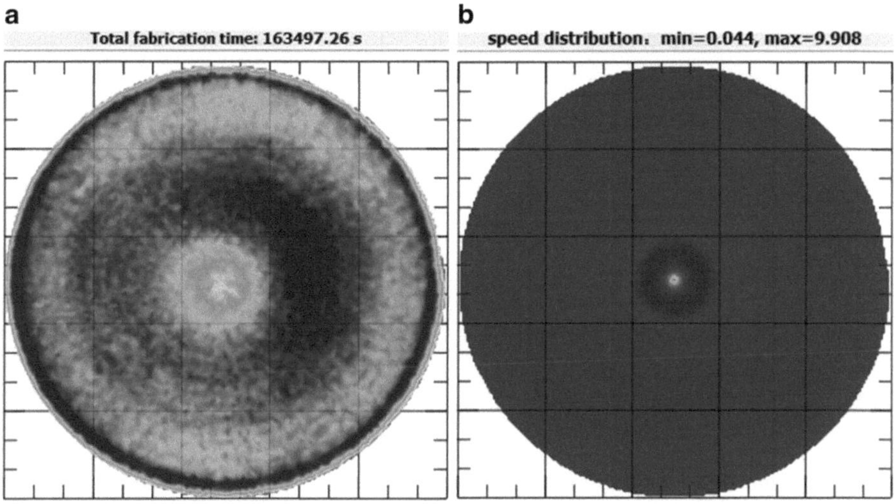

Fig. 4.25 Simulations of adding an extra removal layer: (**a**) the dwell-time map, 45.41 h in total; (**b**) the distribution map of desired velocity, maximum value is 9.908°/s

4.4.2.2 Reducing the Peak Removal Rate PRR of the Tool Influence Functions (TIF) in all Dwell Points

Shown as Eq. (4.30):

$$\mathrm{PRR}' = \alpha \cdot \mathrm{PRR} \tag{4.30}$$

where PRR is the peak removal rate; PRR' denotes the revised PRR; α is the degeneration factor of the TIF.

Similar to the first strategy, it is easy to know that this strategy would globally enlarge the magnitude of the dwell-time map. As a result, the transverse velocity of the polishing tool can be decreased a lot. In the above simulations, its PRR should be changed from $PRR = 1.0\ \lambda/\mathrm{min}$ to $\mathrm{PRR}' = 0.29\ \lambda/\mathrm{min}$, which can reduce the desired transverse velocity to be less than 10°/s.

The FT is increased to 20.88 h. This method also largely increases the total dwell time. Fortunately, the convergence rate of the surface error map would not be decreased. This method can be adopted if the maximum transverse velocity only slightly exceeds the maximum allowed velocity. In addition, the reduction of PRR can be easily achieved by decreasing the tool's spinning velocity, abrasive granularity, and the strength of magnetic field, etc.

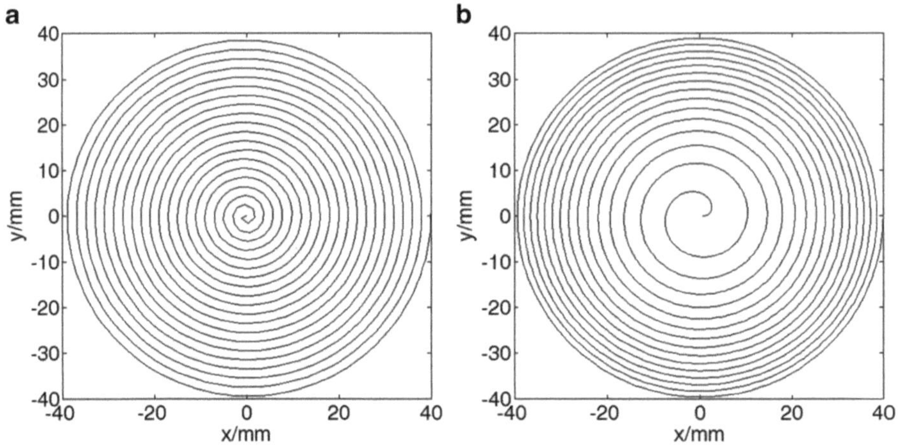

Fig. 4.26 Spiral polishing path: (**a**) uniform distribution: $\beta = 1$; (**b**) non-uniform distribution: $\beta = 0.5$

4.4.2.3 Traversing Non-uniform Polishing Paths

For a polishing run employing polar coordinate, the Archimedes spiral with identical pitch is usually used, as shown in Fig. 4.26. It is denoted as Eq. (4.31):

$$\rho = b \cdot \theta \tag{4.31}$$

Where ρ, θ represents the position of the polishing path in the polar coordinate; b is a constant proportionality coefficient that relates to the screw pitch (screw pitch $= 2\pi b$). It indicates that the screw pitches are equal all over the path.

To reduce the desired velocity in the central region, the dwell time in the central points should be increased by reducing the number of central points. As a method, the screw pitches of a spiral path could be diminished with the increase of radial distance: ρ. This can be defined as Eq. (4.32).

$$\rho = b \cdot \theta^{\beta} \tag{4.32}$$

where β represents a pitch factor. If $\beta > 1$, the screw pitch would increase as ρ increases; $\beta < 1$ denotes that the screw pitch would decrease as ρ increases; $\beta = 1$ is the Archimedes spiral. Set $\beta = 0.5$, the path is plotted in Fig. 4.26b. The path density of the central region is much lower than that of the edge region.

This method can partly decrease the path density in abnormal regions. As a result, the dwell time of these points would be enlarged to decrease the desired transverse velocity. As shown in Fig. 4.27, the tool takes a non-uniform path: $b = 2.94$, $\beta = 0.5$. After simulation process, the results show that PV and RMS are reduced to 0.98 λ and 0.091 λ, respectively. The convergence rate is largely degenerated. This method probably results in insufficient material removal and

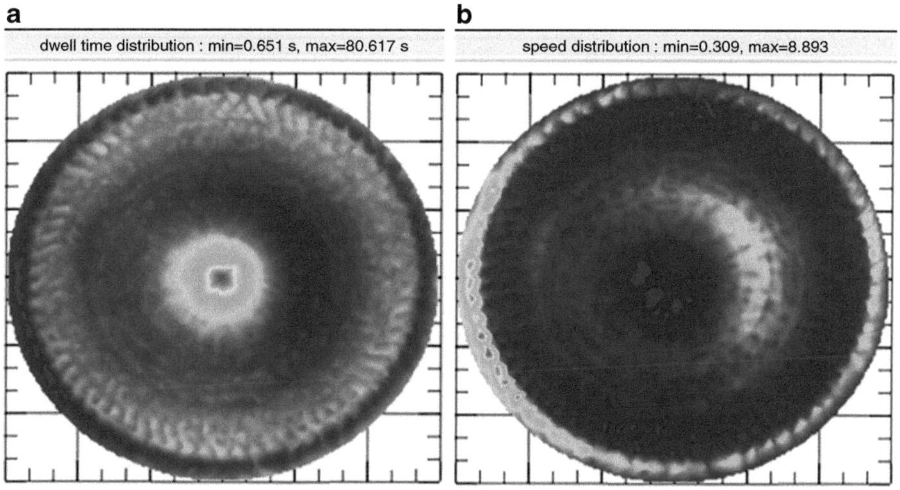

Fig. 4.27 Simulations of non-uniform path: (**a**) the dwell-time map, 6.42 h in total; (**b**) the distribution map of desired velocity, maximum value is 8.893°/s

mid-spatial frequency errors. The maximal transverse velocity stays below 8.893°/s, and the central velocity is obviously diminished. The FT is prolonged to 6.42 h.

4.4.2.4 Reducing the PRR of the TIF Only in Abnormal Dwell Points

When the tool is dwelling in the abnormal points, the factual transverse velocity of the tool or workpiece is less than desired. If we do not change the PRR of the TIF, it would remove more material because of the increase of dwell time. We present a strategy that decreases the spinning velocity of the wheel to reduce the PRR of TIFs for compensating the extra material removal in abnormal dwell points. The wheel spins at a constant velocity in normal dwell points, but slows down when dwelling in abnormal points. This method does not increase the FT, but obtains higher convergence rate. The TIF is time-invariant in normal regions, but time-variant when the tool is dwelling in abnormal regions.

We should illustrate that any method that can diminish the PRR of TIFs could be used to compensate the extra material removal, such as varying magnetic field intensities, plunge depths, fluid fluxes, etc. Considering operability and stability, varying the spinning velocity of the polishing wheel may be preferable. It is much easier to realize by computer numerical control (CNC) technology.

All the above strategies can compensate the implementation errors of the dwell-time map with some conditions. However, the performances are also largely distinct because of the disparity of these approaches. The first and second strategies execute total enlargements. The dwell time of all dwell points is scaled and enlarged, which would largely reduce the fabrication efficiency. Nevertheless,

Table 4.5 Contrast list of four strategies

S/N	CR (%)	FT (hour)	Operability	Versatility
0	90.86	5.97	...	...
1	72.61	45.41	Easy	Weak
2	90.86	20.88	Easy	Weak
3	68.22	6.42	Easy	Weak
4	≈90	6.08	Not so easy	Strong

both the third and fourth strategies are localized processing to abnormal points. They can ensure the fabrication efficiency would not be influenced clearly.

The performances of all strategies with respect to convergence rate, FT, operability, and versatility are listed in Table 4.5, where $S/N = 0$ indicates the theoretical simulation results. $S/N = 1, 2, 3, 4$ represent four strategies. The second and fourth strategies show higher convergence rate than the others. The third strategy has the lowest convergence rate. The third and fourth strategies are much better than the others in FT performance. However, the first and second strategies enlarge the FT multiple times. With respect to operability, the former three strategies are easy to achieve. The fourth strategy needs the support of the FT library. It is especially suitable to commercial mass fabrication. From the view of versatility, the fourth strategy is most powerful, and the former three may suffer a lot of limitations. Based on these performances, we take the fourth strategy as the optimal strategy for improving the implementation accuracy of the dwell-time map.

4.4.3 Compensation Model Establishment and Experimental Validation

The compensation model needs the foreknowledge relationship between the vectors of PRR: PRR_w and the spinning velocity of the polishing wheel: V_w. In this section, we obtained relationship curves through a series of TIF spot experiments with different spinning velocities. The MRF polishing setup equips planetary motion structure [4]. The wheel can rotate along its radial direction (i.e. orbital motion) while spinning to supply rotational symmetric Gaussian-like TIFs. Figure 4.28 concludes the PRRs with the increase of spinning velocity. The upper curve indicates the plunge depth of MR fluid is 1.5 mm, whereas the other is 1.0 mm. The PRR increases non-linear monotonously when the spinning velocity is less than 450 rpm, but decreases slightly when the wheel spins too fast. In practice, the wheel takes 300 rpm and 50 rpm as constant spinning and orbital velocities, respectively. In addition, a useful TIF library can also be built through these experiments.

Assuming that the desired velocity of an abnormal line segment is V_t, the maximum allowed velocity of the turntable is V_m. The constant PRR in the normal line segments is P_n, and normal spinning velocity is V_s. Then, the revised PRR in

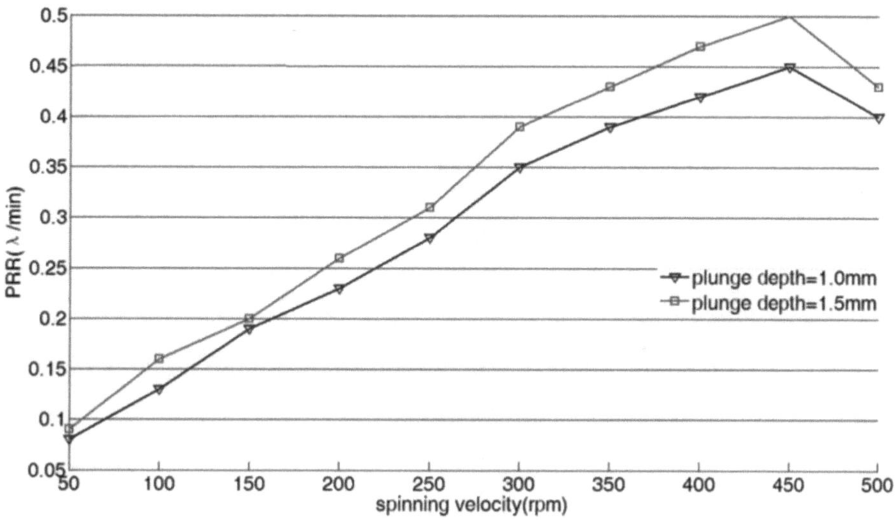

Fig. 4.28 Relationship curves of the PRR of TIFs with respect to spinning velocity of the polishing wheel. *PRR* peak removal rate, *TIFs* tool influence functions

the abnormal line segments P_a could be obtained as Eq. (4.33). Carrying out interpolation operation to the relationship curve of PRRw with respect to spinning velocity V_w, the revised spinning velocity of the polishing wheel V_{sr} could be obtained, as indicated in Eq. (4.34). Here, 'interp' represents the one-dimension interpolation operation, which could be easily achieved in common mathematical software (e.g. MATLAB). During the fabrication process, if the dwell point is normal, the wheel would spin with constant velocity Vs. However, when the wheel dwells in abnormal points, the wheel would slow down to V_{sr}.

$$P_a = V_m / V_t * P_n \tag{4.33}$$

$$V_{sr} = \text{interp}(V_w, \text{PRR}_w, P_a) \tag{4.34}$$

We performed an experiment on a Φ100 mm plane workpiece with initial surface error map: PV = 283.25 nm, RMS = 45.73 nm, as shown in Fig. 4.29a. A spiral path with a screw pitch of 0.8 mm was adopted as the polishing path. The spinning velocity in all points kept 300 rpm; orbital velocity kept 50 rpm; the PRR is 0.38 λ/min; and the TIF covers a region of 8 mm × 8 mm. In this polishing run, the desired velocity of the turntable in the central region (the radial distance less than 7.3 mm) approximately decreases from 55°/s to 15°/s, whereas the maximum velocity allowed of the turntable is 15°/s. The spinning velocity of the wheel is not compensated. As a result, the material removal in the central region was excessive to produce a dent with a maximum depth of 76 nm. As shown in Fig. 4.29b, the residual surface error map has PV = 105.44 nm and RMS = 10.08 nm. It needs

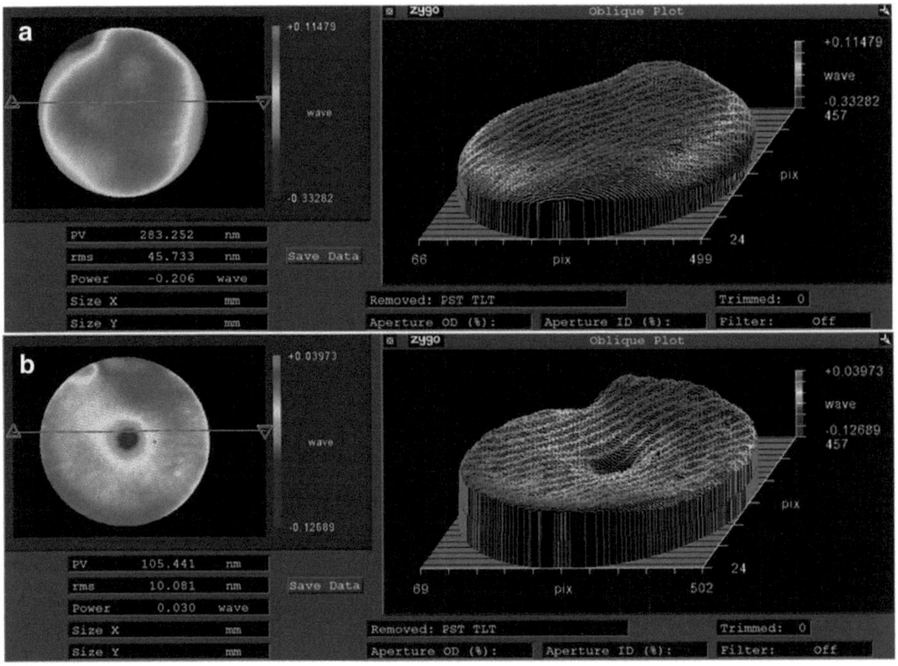

Fig. 4.29 Polishing without compensation model: (**a**) before polishing: PV = 283.25 nm, RMS = 45.73 nm; (**b**) after first polishing run: PV = 105.44 nm, RMS = 10.08 nm. *PV* peak-to-valley, *RMS* root square mean

another polishing process to remove the dent. However, the same velocity limitation problem will still occur. The depth of the central dent may be enlarged if no compensations are adopted.

Based on the above experiment, we conducted the second polishing run to the workpiece. The TIF and polishing path are same as the first polishing run. In this run, the maximum desired transverse velocity reached 304°/s in the central region because of the influence of the central dent. We revised the desired transverse velocity to ensure that the maximum velocity is less than 15°/s. The desired and revised transverse velocity curves are presented in the left *Y*-axis of Fig. 4.30. The spinning velocity curve of the wheel after compensation is plotted in the right *Y*-axis of Fig. 4.30. To contrast obviously, Fig. 4.30 adopts logarithmic coordinate. The horizontal ordinate indicates the number order of the polishing path. The orbital velocity kept constant at 50 rpm. The spinning velocity of the wheel is diminished when the desired velocity is greater than the revised velocity. As shown in Fig. 4.31, after this polishing run, the surface error map shows PV = 38.93 nm and RMS = 4.70 nm. It is easy to find that the central dent was completely removed, which greatly relied on the compensation model

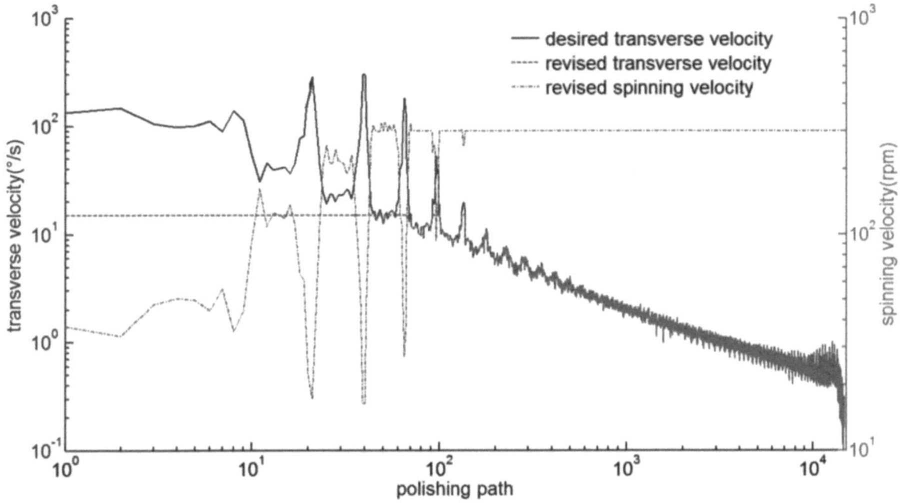

Fig. 4.30 Desired and revised transverse velocity curves of the turntable (*left Y*-axis) and the revised spinning velocity of the polishing wheel (*right Y*-axis)

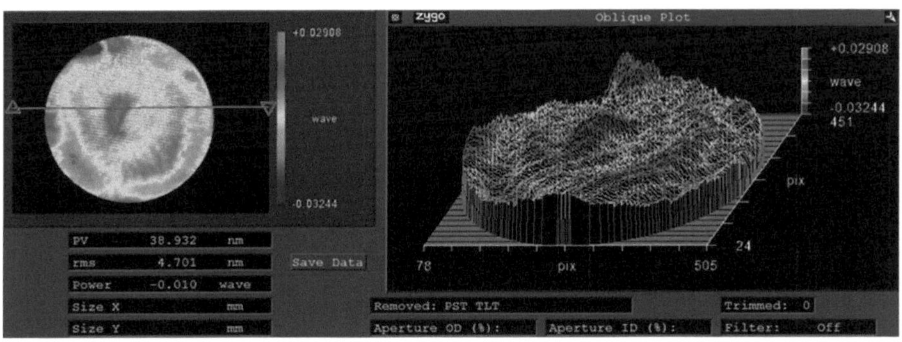

Fig. 4.31 Surface error map after the second polishing run with compensation model: PV = 38.93 nm, RMS = 4.70 nm. *PV* peak-to-valley, *RMS* root square mean

References

1. Jones RA (1977) Optimization of computer controlled polishing. Appl Opt 16(1):218–224
2. Wilson S, McNeil J (1987) Neutral ion beam figuring of large optical surfaces. In: 31st annual technical symposium. International Society for Optics and Photonics, Bellingham, pp 320–324
3. Carnal CL, Egert CM, Hylton KW (1992) Advanced matrix-based algorithm for ion-beam milling of optical components. In: San Diego'92. International Society for Optics and Photonics, Bellingham, pp 54–62
4. Cheng H, Yam Y, Wang Y (2009) Experimentation on MR fluid using a 2-axis wheel tool. J Mater Process Technol 209(12):5254–5261

5. Song C, Dai Y, Peng X (2010) Model and algorithm based on accurate realization of dwell time in magnetorheological finishing. Appl Opt 49(19):3676–3683
6. Deng W-j, Zhang X-j, Wang X-k, Wang X (2009) Novel method for optimizing polishing tool-path in CCOS based on weighted-iterative algorithm. In: 4th international symposium on advanced optical manufacturing and testing technologies: advanced optical manufacturing technologies. International Society for Optics and Photonics, Bellingham, pp 728214–728216

MIX
Papier aus verantwortungsvollen Quellen
Paper from responsible sources
FSC® C105338

If you have any concerns about our products,
you can contact us on
ProductSafety@springernature.com

In case Publisher is established outside the EU,
the EU authorized representative is:
Springer Nature Customer Service Center GmbH
Europaplatz 3, 69115 Heidelberg, Germany

Printed by Libri Plureos GmbH
in Hamburg, Germany